Christian Veder

Landslides
and Their Stabilization

With Contributions by Fritz Hilbert

Translated by Erika Jahn

With 116 Figures

Springer-Verlag
New York   Wien

Christian Veder
Department of Soil Mechanics,
  Rock Mechanics and
  Geotechnical Engineering
Technical University
Graz, Austria

Friedrich Hilbert
Department of Chemical Technology
  and Anorganic Substances
Technical University
Graz, Austria

The cover illustration shows a motorway embankment after a landslide (see Figure 6-40).

Library of Congress Cataloging in Publication Data
Veder, Christian.
  Landslides and their stabilization.
  Translation of: Rutschungen und ihre Sanierung.
  Bibliography: p.
  Includes index.
  1. Soil stabilization. 2. Landslides. 3. Subsidences
(Earth movements). 4. Slopes (Soil mechanics). I.
Hilbert, Friedrich. II. Title.
TA710.V413   624.1′51363   81-5752
                             AACR2

This English language edition is a translation of the Austrian edition, *Rutschungen und ihre Sanierung,* © 1979 by Springer-Verlag/Wien—New York.
ISBN-13:978-3-7091-7606-1

9 8 7 6 5 4 3 2 1

ISBN-13:978-3-7091-7606-1   e-ISBN-13:978-3-7091-7604-7
DOI: 10.1007/978-3-7091-7604-7

Preface

This book was written with the objective of providing geotechnical engineers with a practical guideline on how to cope with landslides as well as of acquainting them with the present state of physical fundamentals and scientific explanations for the phenomenon of landslides.

The book is based on my personal experiences, gathered over decades of work as geotechnical engineer on construction sites in Austria and many other parts of the world, which I also use in my lectures at the Technical University of Graz, Austria.

The method of stabilizing landslides by short-circuit conductors has been developed by myself and has been patented in Germany and Italy. A number of publications already exists (see References) on this method, and of course I also deal in this book with its theoretical and practical aspects.

Here I want to thank my assistants, Messrs. J. Dalmatiner, K. Eigenberger, E. Garber, H. Kienberger, R. Pötscher, and W. Prodinger, for working with me on various projects and for assisting me in the drafting of some chapters of this book, Mr. A. Trippl for preparing the illustrations, and my wife for many a Sunday worked through with me.

I am especially indebted to Dr. F. Hilbert, who has contributed chapters from the field of physical chemistry of clay and silt soils and Dr. J. Gsellmann, who assisted most efficiently in the field measurements. These presentations, dealing with the physicochemical properties of clay minerals and the effects the action of water has upon them, serve to explain the many phenomena to be observed where landslides occur.

Graz, December 1980   Ch. Veder

Contents

Landslides
and Their Stabilization

1 Introduction

This book first describes the main types of slides in soil, their causes and consequences, and then briefly the theoretical basis for stability calculations and the required field and laboratory investigations. This part is intended as a preparation for and introduction to the methods of landslide stabilization described in chapter 6.

For the designing engineer the stabilization of landslides is becoming more and more important, as it is necessary to an increasing degree to construct buildings and transportation routes in areas that were previously avoided on account of their instability. From this results the necessity to take preventive measures in slide-prone areas, in order to avoid later slide movements and thereby grave danger to life and the property of the population.

1.1 Definition

Landslides, especially in soils, are gravitational downward and outward movements of soil masses that may be set off by the liquefying effect of earthquakes. Typically, soils containing minerals with swelling capacity, such as clay, clayey silt, and marl, are frequently fissured and the action of pore-water pressure is especially harmful to stability.

Landslides develop in unstratified, homogeneous soils or are induced by certain types of stratification, e.g., interposed layers of sand or clay. Other causes are excessive load pressures at the head of a slope or undercuttings at the foot. In some cases the soil structure is gradually weakened by chemical or physical processes.

This book deals exclusively with the stabilization of landslides in soils; the following types of slides are *expressly* excluded from the considerations:

(1) Debris avalanches (the German term is *Mure,* also *Murgänge* or *Murbrüche*), as they occur in mountainous regions, where after heavy rains or sudden melting a mixture of water, earth, and rubble decends from steep slopes and in mountain torrents with devastating effects, such as the damming up of rivers or the destruction and burying of settlements and transportation routes.

How to prevent or stop such debris avalanches is the responsibility of local authorities in charge of mountain torrent regulation.

(2) Rock slides or rock falls, where tremendous rock masses, sometimes several million cubic meters, are loosened, fall down and, after they hit the ground below, descend farther in an avalanche of fractured rock. How to prevent and stabilize such rock falls lies in the field of rock mechanics.

(3) Rock slides in clayey shale, phyllite, and mica. The stabilization of such slides is discussed only if the rock has a similarity to soil. Genuine rock slides are treated by detail by Zaruba and Menzl (1969) and Müller (1964).

1.2   Economic Significance of Landslide Stabilization

Almost weekly we hear about landslides with more or less devastating consequences for life and property of the population in the affected area. Stabilization measures need to be adapted to the type of landslide. It requires a high degree of experience and flexibility on the part of designer and contractor, first to make the right selection among the often quite expensive stabilization constructions and secondly, to react immediately and on the spot when it becomes apparent, from new factors during construction work, that the original design needs to be adapted to changed requirements. In selecting the safety coefficient, which is of greatest importance in any construction project, the designing engineer must apply different measures for the stabilization of landslides from those that he normally would apply for other types of construction. To assess the quotient between resisting and driving forces requires a high degree of experience and know-how.

There are instances where it takes some courage to accept a safety coefficient considerably lower than usual in order not to arrive at a design that is too expensive. High stabilization costs could render it uneconomical to carry out a construction project and make it preferable to leave the slide-prone area, either by abandoning the project altogether or by relocating it (e.g., redesigning transportation routes to pass through more stable areas).

A sound design is based on careful observations for possible sites of soil movement before, during, and even after carrying out stabilization measures. Indispensable geological, hydrological, and soil-mechanical investigations also serve this purpose. Besides purely economic considerations, such as the cost–benefit ratio, nowadays social and ecological aspects also play an important part.

The authorities who have to decide whether an area endangered or devastated by landslides is to be saved or evacuated are again and again confronted with the fact that the population, even after severe losses of life and property, refuses to leave the homeland. Immediately following the catastrophe, they will start to rebuild on the same spot, though they know that such phenomena of nature may reoccur. This kind of behavior is demonstrated by the people living for generations on the slopes of volcanoes belching fire and lava; they literally stick to their soil. In such cases purely economic considerations do not make sense and can only lead to grave social and political tensions.

As far as landslides are concerned, *Japan* represents an extreme case. In view of its 7,000 landslides, landslide stabilization has become a national concern. Other countries may be faced by occasional occurrences, a slide here and a slide there, that usually can be handled by local authorities, but Japan must meet this menace by all available means unless the country wants to resign itself to severe damages occurring with increasing frequency.

It is well known that only about 20% of Japan can be inhabited and cultivated; the remaining 80% are densely forested mountains. These figures must be seen in the context of a population of about 115 million. Understandably, every single square meter of soil must be used for the traditional cultivation of rice. But the areas that are especially fertile owing to ample presence of water, are the same ones where water, which is such a blessing for rice crops, has a disastrous effect upon soil stability, generating landslides by the well-known mechanisms.

The Japanese tried in vain to cope with the landslides by primitive methods, such as surface drainage, but lately the situation has become even more critical. Water storage reservoirs for agricultural purposes and power plants, construction of motorways, and irrigation and drainage of large areas by modern methods may cause water movements that trigger slides in areas where none occurred before. Now a carefully orchestrated attempt is being made to gain control. The following institutions work on landslide research and stabilization:

(1) Japanese Landslides Society, consisting of the following members:
   (a) Researchers in the field of geology, geography, and geophysics from construction engineering, forestry, and agriculture, who work at universities.
   (b) Researchers and engineers who are charged by the Ministry of Construction Agriculture and Forestry as well as the Japanese National Railways and the provincial governments with soil investigation and stabilization work.
   (c) Engineers who work for major private companies, construction firms and manufacturers of measuring devices for soil movements.
   The society has 1,200 members, issues a quarterly paper and periodically organizes meetings, seminars and excursions, etc.
(2) National Conference for the Stabilization of Landslides. This is an association of the prefectures of 44 provinces, with the purpose of promoting an exchange of experiences regarding landslide research and stabilization.
(3) Universities. About 40 researchers work at universities to:
   (a) Study the parameters that should make it possible to predict landslides and to analyze the correlation between these parameters and the occurrence of landslides.

        (b)   Develop a theory that makes it possible to predict which changes a natural slope will undergo if the water balance is changed by human activities, i.e., different types of constructions.

(4)   Government. The following departments take part in landslide research and stabilization:

        (a)   Ministry of Agriculture and Forestry—to preserve and protect arable land and forests by stabilization measures.

        (b)   Ministry of Construction—to protect rivers, dams, motorways, residential areas, and public facilities by stabilization measures.

        (c)   Japanese National Railways—to safeguard the operation of railway facilities.

(5)   Research Institutes.

        (a)   National Institute for Agriculture and Forestry Research (attached to the Ministry of Agriculture and Forestry).

        (b)   Geological Institute.

        (c)   Institute for Public Constructions (attached to the Ministry of Construction).

        (d)   Geographical Institute (also attached to the Ministry of Construction).

        (e)   National Research Center for the Prevention of Catastrophes.

        (f)   Technical Research Center of the Japanese National Railways.

        (g)   Technical Research Center of Power Plants.

(6)   Associated Academic Organizations:

        (a)   Japanese Society for Mountain Torrent Regulation

        (b)   Japanese Society of Geotechnical Engineers

        (c)   Japanese Society of Construction Engineers

        (d)   Japanese Geological Society

        (e)   Japanese Geographical Society

        (f)   Japanese Society for Forestry

The Ministry of Construction has spent the following amounts on the observation and stabilization of landslides: in 1952, 389.0 million yen, and in 1972, 5,340.0 million yen; the Ministry of Agriculture and Forestry spent 222.7 million yen in 1952 and 4,972.0 million yen in 1972. This serves to show what a prominent part landslides play in Japan and that only a dedicated mobilization of all scientific and financial resources will serve to bring ·about the vital success.

This gigantic program to combat landslides may also provide other less afflicted countries with suggestions for actions to be taken.

2 Characteristic Types of Slides and Alternatives for Their Stabilization

As may be seen from the literature, there are many ways to classify landslides by types, but only a classification based on the nature of the moving subsoil can serve the purpose, especially in connection with the alternatives for slide stabilization (Varnes 1972, 1978). Geotechnical engineers know that soil as such is a two-phase or three-phase system and that in addition to the solid components, water or air play a dominating part. When confronted with a landslide, the experienced engineer will first investigate the effect of water, which may by present as pore water or ground water, or be mixed mechanically with the soil particles, causing incipient or already-progressing soil movement.

Soils are very rarely truly homogeneous or, to use another word, "unstratified." Also, chemically homogeneous soils, especially if they are overconsolidated, show fissures and cracks that represent weak spots in its structure, and consequently cannot be considered unstratified. But the majority of landslides occur in stratified soils. These are:

A— Loose, water-bearing sand
A_1—Mixtures of gravel, sand, and silt, loosely deposited on stiff clay
B—Soft, fissured clays
C—Stiff, fissured clays
D—Clays with extended planar or pocket-shaped embedments of sand or silt.
B and C may be found in extended layers, whereby the lower layer has different chemical properties from those of the upper one; frequently, the upper layer originates from the decomposition of the lower one. A water film may be present between the clay layers.

The engineer designing the stabilization measures faces a great challenge. Based on observations of the ground surface and the mostly rather meager results from soil explorations in boreholes and trenches, he must be able to take into account the almost infinite variety of combinations of soil properties and hydraulic conditions. In addition, the designing engineer must be aware of the construction problems of the various stabilization methods and be able to evaluate these properly. Lastly, he must submit estimates regarding duration and costs of construction work, and here it is advantageous if he can draw on personal experience in general, but also on a knowledge of local conditions. Chapters 3, 4, and 5 cover significant aspects to be taken into account by the designer before he begins to work out stabilization measures.

The following chapters deal with details which—as the title of this book indicates—are of significance in the stabilization of landslides. The description and discussion of soils and slide phenomena, where to my knowledge no stabilization efforts have been undertaken so far, are omitted. Without claiming completeness, I have compiled a number of stabilization alternatives in Chapter 6, most of them based on my designs.

3 Main Causes of Landslides

3.1 Geological Causes

As already explained in chapter 2, this book, without claiming completeness, deals primarily with landslides in stratified soils. With only a few exceptions, these are soils from the younger Tertiary; they are frequently marine sediments originating from the Sarmate (uppermost Miocene) and the Pont-Pannon (lowest Pliocene).

Usually these deposits consist of layers of clayey silt, sandy silt, and silty sand. The thickness of the layers varies between a few centimeters and several meters. The stratified structure of sedimentation may be explained as follows: Depending upon their water volume and velocity of flow, rivers discharging into the sea carried along larger or smaller quantities of fine particles that were deposited on the ocean floor. This sedimentation took place either evenly or irregularly, being influenced by the changing directions of marine currents. Frequently the layers are gently inclined (about 5° to 10°); in part this may be attributed to oblique fluvial deposits during delta formation, and in part to orogenetic effects.

The layers are usually water-bearing and it may be observed that frequently water is present either as a thin film at the zone of division between clayey and sandy silt or as a laminary flow in more permeable sand layers. It usually flows out at the surface where the sand layer is wedging out, which happens when the slope has a steeper inclination than the sand layer.

In relatively thin sand layers a considerable water pressure may build up if more water enters from the top than is discharged at the outlet (see 6.1.4.1 (II)). This perched water is one of the main causes of landslides. Thicker sand layers rarely are fully water-saturated and may even serve as natural drainage.

Under certain circumstances tectonic movements have led to the formation of marked slickensides in the layers (see 6.1.3). If these slickensides have an unfavorable inclination, for example towards the foundation excavation, then sudden slope movements may set in even in the absence of water; these movements often run transverse to old slip surfaces, mostly with a marked inclination towards the bedding planes.

Frequently, for instance in the area around Graz, Austria, the layers were under a previous load pressure from several 100-m high gravel and sand masses. This load pressure causes clayey silt layers to settle less than could be expected from their consistency or plasticity. Sandy silt layers are compressed to a degree that renders them quite compact and stable as long as they are not disturbed by construction equipment; however, they will promptly change into a completely unstable, gritty slush the moment that they are disturbed and mixed with water. Also, a minimal but nonetheless slide-producing water film may be observed at the zone of division between layers. A special type of stratification (solifluction) is described in 3.5.4.

In Japan, for instance, a marked difference may be observed between marine deposits, such as marl, and soils of volcanic origin, such as tuffs (these originate in the Miocene; see 6.3, Table 6-2).

3.2 Morphological Causes

The formation of the terrain and its inclination play an important part in the stability of a slide area. By experience it may be stated that slopes with a 2:3 inclination (vertical to horizontal) generally are stable. For over a hundred years this inclination was considered acceptable when planning railway or motorway routes. As a rule, it is human beings who disturb the equilibrium of slopes by cutting through them and constructing embankments and bridges for transportation routes. But landslides may also occur without human action, from erosion (undercutting at the toe by unregulated rivers) or from debris deposits (overloading the head of the slope).

3.2.1 Excessive Steepening of Slope Inclinations

Regarding a landslide that originated at the zone of division between overlying brown and subjacent blue clay, Skempton (1977a) cites, among others, the example of a 23-m deep cut (New Cross Cutting) near London, where 3 years after the completion of construction work a slope with a 2:3 inclination (vertical to horizontal) slipped down suddenly and without warning; the moving mass amounted to about 40,000 m³.

As another example, Skempton (1977b) describes the cut at Potters Bar, between London and York, where in 1850 an inclination of 1:3 was chosen initially. When the cut was widened in 1956, the inclination was leveled down to 1:4 at the foot for reasons of safety; the head was kept at 1:3. Skempton and Hutchison (1969) further describe three landslides at a cut near Bradwell, Essex, where the inclinations chosen were (from bottom to top) 2:1, 1:1, 1:1, with intermediate berms 2 and 4 m wide. The foot of these slides was quite close to the zone of division between blue and brown horizontal clay layers. The clay had the following soil-mechanical characteristics: $w = 33\%$, $w_L = 95\%$, $w_p = 30\%$; it was under a previous load pressure from about 170 m of sediments.

These slides started 5 to 19 days after excavation work was completed, but cases are known where landslides occurred 15 to 40 years later. According to Skempton (1977c) this is due to the fact that the value $\bar{r}_u = \gamma_w \cdot h/\gamma \cdot z$ slowly builds up in the course of time, i.e., 40 to 60 years (γ_w = unit weight of water, γ = unit weight of soil, h = piezometric height above slip surface, z = height of ground surface above slip surface). The only efficient stabilization measures are to level down the slope or to construct anchored walls (see 6.2.9.4).

3.2.2 Excessive Load Pressure on Slope Head

An excessive load pressure on the upper section (head) of a slope may lead to a landslide. In 6.2.1.2 the example of an overloaded abutment as insufficient support for a major embankment fill is presented. Strong soil movements occurred and the abutment was destroyed so that the fill had to be leveled down and the abutment set back.

6.1.2 (II) describes the movement of a slope that had an inclination of only 6° to 9°, set off by the fill for a motorway embankment. Here the movement was stopped by 28 wells (3 m in diameter) installed at center distances of 9 m and reaching down to depths of 20 m.

3.2.3 Weakening of Slope Toe

The weakening of the toe of a slope may have several causes.

(1) Flowing water may erode and undercut the toe of a slope (see 3.5.3.3). In this case the best measure is to protect the toe, either by a carefully carried out rock fill or by a wall, unless it is possible to divert the river. As a preliminary measure, any building situated on the slope may be protected by a supporting construction until the slope is securely stabilized, as described in 6.1.1 (I). Such a measure, however, is only effective over a limited period of time, since the part of the slope below the supporting construction becomes steeper and steeper as undercutting progresses, and limit equilibrium is bound to be reached. At this point the slope slides down and exposes the support, and this may have catastrophic consequences for the building.

(2) The toe of a slope moves downward through the action of water streaming in the subsoil (see 3.5.2.2). From the theoretical explanations in 4.1, Example 3, it may be seen that, as a rule, a flow of ground water moving parallel to the slope reduces stability by 50% as compared to a slope without ground-water flow. Such a case is presented in 6.2.3. To stabilize the slope, the loose material at the toe was replaced by a stone wedge; this lowered the ground-water table and at the same time considerably increased shear resistance.

(3) Present-day design of motorways aims at a smooth line and frequently this makes it necessary to cut through steep slopes that are almost at limit equilibrium. Before the introduction of modern construction methods, as

described below, such cuts had to be supported by retaining walls. Construction work started with the cut, which had to be carefully braced to hold loose soil in place and to prevent movements in the often very steep and fissured uphill part of the slope—which sometimes also carried constructions. This bracing is very expensive, especially for high cuts, and often it is impossible to provide it with the firm footing required to take up horizontal earth pressures. The wall was constructed as either a gravity or cantilever retaining wall; the bracing was removed simultaneously. The dimensions of these walls in proportion to their height, especially at the foot, required huge additional excavations as compared to modern methods.

An additional disadvantage of retaining walls is that the construction of the bracing, its step-by-step removal during the construction of the wall as well as the fitting of the wall against the often rather loosely deposited natural soil, necessarily has a relaxing effect, sometimes very deep-reaching, within the slope that causes or may cause cracks in the terrain or in buildings.

Krainerwalls, gabionades, reinforced earth, and the new "Ebenseer" wall are methods that for reasons of economy are much in use today; they are less expensive to construct than regular walls but require about the same amount of excavation. Their main advantages are their flexibility where settlements are irregular and the ensured drainage behind the wall (see 6.2.9.5 (I–IV)).

Wherever only limited space is available, the following construction methods are in use today:

— Before the excavation for the motorway or railway route is carried out, reinforced concrete is installed in the natural soil without relaxing it. The retaining wall, if not anchored, is carried out with such a high bending resistance that it requires no support after the excavation. This bending resistance is achieved best by constructing the wall with T-shaped diaphragm-wall panels or by prestressing it vertically.

— If anchoring is foreseen, then diaphragm-wall or bore-pile panels are installed prior to excavation. As excavation work progresses, these wall panels are fixed with one or several rows of anchors. They are virtually strain resistant and take over the function of retaining walls (see 6.2.9.3).

— Where bore-pile or diaphragm-wall panels cannot be constructed, either because there is no space for the heavy construction equipment or because this method cannot be used owing to the presence of large, very hard boulders that are difficult to intersect, an anchor wall is the economical and technically sound solution frequently chosen (see 6.2.9.4). This type of wall is constructed from top to bottom by exca-vating 2.5-m high strips, concreting the wall section and fixing it to the natural soil with built-in anchors. The soil is not relaxed and the slope remains completely at rest. This construction method requires the least space as no heavy machinery needs to be used.

— Another alternative that is still in the experimental stage is the nailing of the soil (see 6.2.9.5 (V)).

3.3 Physical Causes

3.3.1 Decay of Cohesion with Time

In this section cohesion c' of the drained soil is treated (see 3.3.3, Fig. 3-1b), as it exists in overconsolidated cohesive soils. A pseudocohesion is briefly mentioned first, i.e. moist sand is held together by the surface tension of meniscal water and this cohesion disappears the moment the sand is soaked or dries out.

In the course of the history of the Earth, clay wash was sedimented and then consolidated under a growing load. At a later stage of geological development, these deposits were removed and overconsolidated clay with a distinct cohesion remained. Overconsolidated clays are further distinguished by their ability to regain deformability, depending upon the consolidation pressure and increasing with their plasticity (Cheney and Fragaszy, 1980). After the load is removed, this clay expands in volume as it regains deformability. Normally a volume expansion in the direction of the surface is possible, but expansion parallel to the surface is confined, and this leads to a concentration of stress parallel to the surface. According to Skempton (1961), the coefficient of earth pressure at rest k_0 may reach values above 1.0. Locally, stress conditions may exceed the shear strength of the clay and this leads to the development of cracks, which represent the first disruption of cohesion. Water from precipitation enters the cracks, the water content increases, and the surrounding soil becomes soft (Crozier, 1969; Nishida et al., 1979). This process further decreases shear strength. In addition, the swelling process generates new stress conditions and new cracks may develop. This mechanism is described in more detail in 3.4.4.

The uppermost soil layer is most exposed to weathering. Changing temperatures and alternating soaking and drying disturb the soil to a degree that it eventually loses its cohesion altogether. In addition, it undergoes chemical processes, such as oxidation and decomposition. The subjacent zones are also under such influences, namely seasonal fluctuations of the ground-water table (pore-water pressure) and changes in temperature. These influences, together with gravitational forces and the stress relaxation parallel to the surface, may cause slope movements that reduce shear strength, primarily cohesion c', but also the angle of internal friction φ'. In this connection reference is made to 3.3.3 (progressive failure).

Stratified soils are especially susceptible to weathering if the layers crop out with more or less of a gradient. Here meteoric water penetrates the fissures of the layered packs, dilates them under the influence of frost and softens the soil, so that the structure of the mass as a whole is disturbed and its strength decreased (Crozier, 1969). This applies also to rocky soils, such as sandstone, marl, and clay slate. A relevant stabilization example is presented in 6.1.4.1 (I). When faced with more silty soils, an attempt can be made to stabilize the slope material with lime (see 6.1.5.3). Furthermore, clay soils can be drained by the electroosmotic method after Casagrande (see 6.1.5.2), at the same time increasing the angle of internal friction by introducing Al-ions from the anode, thereby strengthening the soil structure (electroosmotic injection).

Here it should also be pointed out that a lot of time may pass before cohesion decreases to a degree that soil or rock slides set in. In 1805, for example, a rockslide of this type destroyed the village Goldau in Switzerland. The mass of Nagelfluh, descending on a slip surface, had a volume of about 15 million m³ (Redlich et al., 1929). The cause for this slide was found in two interacting factors: One was the decrease of cohesion in the bedding plane due to weathering (frost, leaching) plus a relaxation process in the valley slopes, the other the building up of the water pressure in the bedding plane.

Rockslides especially, but also soilslides of such proportions, are actually not preventable and it is hardly possible to predict when the catastrophe will occur, as resources for permanent observation of such large areas by strain gauges or geological surveys are generally not available.

3.3.2 Diagenetic Cohesion and Its Decay as Related to the Danger of Landslides

As an introduction to this chapter, a short description of the main characteristics of diagenesis is given. Diagenesis is a dehydration and solidification of the soil going beyond the common consolidation process, i.e., a preconsolidation due to a certain height of superimposed deposits. Sometimes solidification takes place by way of cementation and minor chemical reactions. Diagenesis comprises, for example, the solidification of clay until it becomes mudstone and the formation of sandstone and conglomerates, or breccia. Temperatures up to 260° to 270°C are involved in the process. With higher temperatures one speaks of metamorphosis. Under the pressure of later deposits, granulation, recrystallization, an exchange of substances (metasomatic reaction), and the formation of new minerals may take place in the diagenetic zone, as far as this is possible in sedimentations.

The following example of a consolidation and diagenetic solidification is limited to the formation of sedimentary rock from claywash. After sedimentation, under the growing load pressure from later deposits and through diagenetic solidification, clay and mudstone form. At a later stage of diagenesis, the stratification originating from the sedimentation process may resemble a schistosity and for this reason one speaks of clayslate and schist.

The properties of diagenetically solidified clays have been treated in detail by Bjerrum (1967) and here only the significant characteristics are pointed out. The regainable deformability of overconsolidated clays has been mentioned in 3.3.1. Depending upon the magnitude of the diagenetic cohesion (depending in turn upon the mineral constituents), this ability to regain deformability is more or less blocked in diagenetically solidified clays. Therefore, clays with a strong diagenetic cohesion, when relieved of load pressure, show a very minor tendency to expand in volume and consequently exert only a slightly increased lateral pressure. But even if the load pressure is increased beyond the original solidification pressure, the deformation in a particular zone will be minor owing to the diagenetic cohesion. Contrary to common overconsolidated clays, it is therefore not possible to draw exact conclusions regarding the overconsolidation pressure from the stress-strain diagram.

When diagenetically solidified clays are relieved of load pressure and exposed to weathering by the removal of superimpositions, the retarding effect of the diagentic cohesion, as compared to normal overconsolidated clays, sets in. However, in the course of time the diagenetic cohesion may disappear through weathering and relaxation, so that in the end the properties of these clays may be similar to those of overconsolidated clays. Therefore the dangerous aspect of clays with diagenetic cohesion lies in this retarding effect which makes it difficult to assess a critical situation. Slopes may remain stable over decades or longer and then suddenly lose their stability. Chapter 3.3.3 discusses this phenomenon again.

3.3.3 Progressive Failure—Engineering Geology of Overconsolidated Plastic Clays

Frequently slides occur even though the shear strength required for the stability of the slope is below peak strength. This may be explained by the presence of a weak zone in the slope where, through movements, peak strength was surpassed and shear strength decreased below peak strength, in extreme cases even down to the residual strength. In such weak zones progressive failures may start, whereby the slip surface will progress from bottom to top, in the direction opposite to the soil movement.

The following deals with the preconditions for the formation of progressive failures.

(A) A progressive failure can occur only if the soil shows a marked difference between peak strength τ_f and residual strength τ_r and only if the mean shear strength required for the stability of the slope is larger or only insignificantly smaller than τ_r (Fig. 3-1). Furthermore, it is easily understood that the susceptibility to progressive landslides becomes the larger, the smaller the displacement v required to reduce shear strength from τ_f to τ_r. On account of the required large difference between τ_f and τ_r, usually only strongly overconsolidated soils, primarily clays, silty clays, clayslate, and schist, are prone to develop progressive failures.

In Oslo in 1967, Bishop introduced the brittleness index $I_B = (\tau_f - \tau_r)/\tau_f$. The greater the brittleness of a soil, the greater its tendency to develop progressive failures.

(B) As mentioned at the beginning of this chapter, another precondition is the presence of a weak zone in the slope. Such weak zones cause a concentration of shear stresses. In addition, conditions for a differentiated deformation

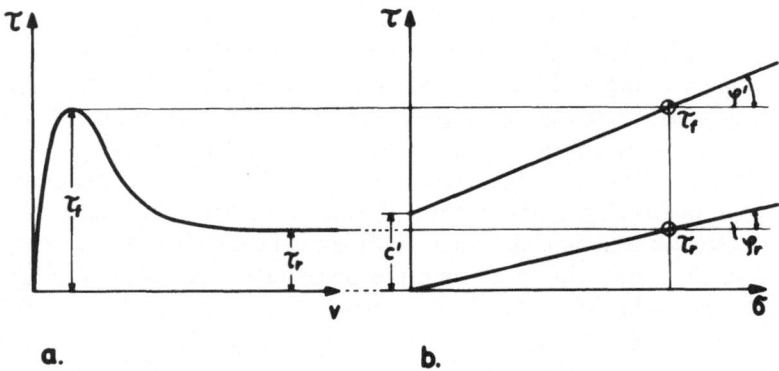

Figure 3-1 Shear strength of overconsolidated clay; a—in correlation to displacement v; b—in correlation to normal stress σ.

must prevail. There are various possibilities for the development of a weak zone, such as man-made cuts or simply an undercutting of the slope foot by erosion; the shear resistance of a soil may also be decreased in places by weathering and subsequent softening (3.3.1), or the diagenetic cohesion may decay, also through weathering (3.3.2). As the diagenetic cohesion decreases, the soil tries to expand but normally can only do so in the direction of the slope surface, since the lateral expansion is confined. The consequence is stress parallel to the slope that may cause a local failure. For a more detailed study of the process connected with a progressive failure, Bjerrum (1967) and Bernander (1978) are recommended.

In a simplified model of a uniform, stable slope with gradient β (4.1, Example 5), shear stress τ_0 acting on a plane parallel to the slope at depth h, being smaller than peak strength τ_f, stems solely from gravity. The earth pressures acting at an assumed lamella are opposing and of equal magnitude. In order to set off a progressive failure, there must be a point of weakness, as already outlined above, assumed in Fig. 3-2a to be a vertical excavation down to h. This activity disturbs equilibrium insofar as the original lateral earth pressure, which is still effective at a certain distance from the excavation, must be taken up at A on account of an increase in shear stress in the plane parallel to the slope. This inevitable distribution of shear stress, with the maximum value at A, will take approximately the course shown in Fig. 3-2b. However, if maximum stress, as assumed in Fig. 3-2b, exceeds the shear strength of the soil, then movements set in which reduce shear strength, as shown in Fig. 3-1a, down to residual strength, since a sufficiently large deformation has been assumed in this model. Owing to this decreased shear strength adjacent to the excavation, the stress concentration moves uphill, as shown in Fig. 3-2c, and there again leads to a surpassing of peak strength and a reduction of residual strength, and so forth.

Depending upon gradient conditions in a slope and the resulting gravita-

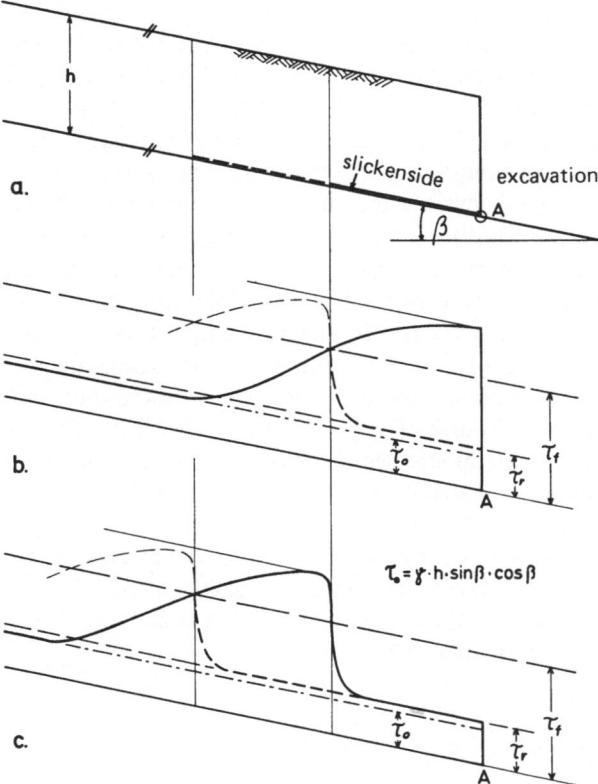

Figure 3-2 Development of progressive failure—a; b and c show the first and second phase of the process.

tional shear stress $\tau_0 = \gamma \cdot h \cdot \sin \beta \cdot \cos \beta$ in relation to residual strength τ_r, there are three possible variations:

(1) $\tau_0 < \tau_r$: Since the residual strength τ_r is larger than τ_0, the stress concentration decreases with the distance from the point of original disturbance because the lateral earth pressure active in the undisturbed zone is sustained by the subjacent soil. Where the maximum value of required shear stress no longer exceeds τ_f, no further slope movements take place, i.e., progressive failure stops.

(2) $\tau_0 \simeq \tau_r$: Here the process of progressive failure continues practically *ad infinitum;* it will not stop of its own accord. Furthermore, any minor additional disturbance may cause the whole slope to come down.

(3) $\tau_0 > \tau_r$: In this case soil masses with shear strength reduced to τ_r will immediately slide down in slabs. In a uniform slope there will be no state of rest, meaning that the process of progressive failure never stops. The size of the slabs is determined by the cohesive strength of the soil.

Here a possible delaying effect should be mentioned briefly, which may affect the development of a progressive failure and which makes it rather difficult to assess the danger of a sudden slide of large soil masses (unpredictability). In this respect the most dangerous soils are those with a strong diagenetic cohesion. As they are capable of regaining their deformability, overconsolidated clay soils tend to expand when relieved of load pressure. This expansion normally is not confined towards the surface and consequently in such soils, in their original state, horizontal stresses will be larger than vertical stresses. Skempton (1961) reports, for example, that in London clay the coefficient of earth pressure at rest exceeded 2.0 in the uppermost 15 m. The overconsolidation of London clay is due to a former superimposition of a mass at least 150 m high (Skempton, 1961 and 1977a). In soils with a strong diagenetic cohesion the ability to regain deformability is somehow arrested, but may be released by weathering in the course of years and decades (3.3.2). The consequence is a large increase in lateral stresses that may trigger progressive failures in formerly stable slopes or embankments (Maugeri, 1978; Janbu et al., 1977; Morgenstern, 1977).

A kind of retarding effect may also be observed at cuts, where an initially existing pore-water underpressure adjusts to changed conditions. However, this should not be mistaken for the retarding effect described in connection with diagenetic cohesion (Bishop, 1957).

3.3.4 Landslides in Stiff, Fissured Clay

The development of cracks and fissures has already been described briefly in 3.3.1. Their main cause is the relief from load pressure through the removal of superimposed material (glaciers, moraines, alluvions) and other sediments. Owing to its regainable deformability, the clay expands but cannot do so in the direction parallel to the surface. Through this confinement of the deformation process, perhaps also assisted by gravitational stresses due to the slope gradient, the lateral pressure increases to a degree where it surpasses in places the shear strength of the clay, and small movements take place. As a consequence, virtually any type of clay is weakened by fine fissures and slickensides.

If distance between fissures is in the range of centimeters, then the slope may become unstable and slide down already during excavation work or shortly after its completion. Such slides of narrowly fissured clays take place as soon as the shear stress exceeds the average shear strength of the material. They may occur also in man-made slopes with an inclination of only 1:3 (vertical to horizontal) and continue for years.

If the distance between fissures and cracks is in the range of decimeters, then a slope may remain stable for years and even decades after its completion. There are cases where a landslide occurred 80 years after the excavation of a slope with a 1:2.5 inclination, although no signs of streaming water or springs could be observed. After its completion the slope soil loses its strength from the effects of weathering. Surface water enters the original fissures, which are opened up further as the load pressure is relieved in the course of excavation work, and softens the clay surrounding the fissures. Depending upon the degree of soaking and the mineral content, the developing swelling pressures may not be uniform and may break up the soil into single, irregular blocks, which in turn continue to disintegrate until a soft matrix remains containing only single firm pieces which no longer have any stabilizing effect upon the mass as a whole. Such masses then are capable of creep movements similar to those of a viscous liquid (Blight, 1977; Gudehus *et al.*, 1976; Ter-Stepanian, 1974).

A landslide will develop at the toe of a slope as soon as driving forces exceed resisting forces with the average shear strength. Terzaghi and Peck (1967) report on cases where the average shear strength decreased from high values at the time the cut was carried out down to values between 0.02 and 0.035 MPa. Below the surface, however, clay quickly becomes more resistant.

Apparently, the action of water in firm, fissured clays mainly causes a weakening of the clay structure and, thereby, a weakening of its shear strength; only rarely has it been observed that streaming water has an influence upon stability (3.4).

The following stabilization measures may be taken into consideration:

(1)   If an acceleration of slope movements is registered with the aid of benchmarks on the slope surface, then the slope must be immediately graded down as far as possible.

(2)   The head of the slope must be drained as much as possible by surface and subsurface drainage (6.1.4.1 (I)).

(3)   The slope surface must be made as impermeable as possible by filling up the largest cracks, spreading of humus, and planting.

(4)   In some cases a solidification by grouting, the Claquage method, will be indicated (6.1.5.1).

(5)   In other cases, stone wedges or stone ribs may serve to stabilize a slope, provided these reach into unweathered clay and have stone or concrete wall footings. During construction, care must be taken that no clay enters the macropores of the stone ribs; this can be avoided, for example, by the use of filter fabric wrappings (6.2.3). If not too much ground water is present, the stone ribs may also serve as drainage; however, sometimes it is necessary to construct weeping drains with filtering pipes (6.2.3 and 6.2.7).

Measures similar to those described above are also effective in soils with diagenetic cohesion or with a susceptibility to progressive failure.

3.3.5 Effects of Earthquakes

On account of their acceleration effect, earthquakes may trigger not only devastating landslides but also rock falls and the like (Hirao and Okubo, 1971; Newmark, 1965; Seed, 1968; Morton, 1971; Coulter and Migliaccio, 1966). Cotecchia and Melidoro (1974) describe how a series of landslides was triggered by an earthquake in Calabria, which in turn dammed up rivers until they became lakes. However, a different effect of earthquakes is the main subject here: so-called liquefaction (Casagrande, 1976).

Loosely deposited, fully saturated sand (void ratio larger than critical void ratio) may be compacted by seismic tremors (contractant deformation) so that the increased pore-water pressure practically balances the effective stress and the soil liquefies (Seed and Peacock, 1971). Grave damages may be the consequence; constructions, like buildings and embankments, literally vanish into the ground, pipelines break, etc. A way to prevent this liquefaction is the treatment of the soil by vibro-flotation, i.e., the shocks of the vibrator cause the soil to collapse so that the void ratio decreases below the critical limit and no dangerous pore-water pressures can build up through eventual earth tremors. Loess soils, in which the voids are filled by air instead of water, are also able to liquefy under the impact of earth tremors as air pressure instead of water pressure builds up.

Casagrande and Shannon (1947/48) report on a large-scale loess slide in the Chinese province Kansu, triggered by an earthquake on December 16, 1920, that took 100,000 lives. The reports by Close and McCormick (see Casagrande and Shannon, 1947/48) ought to convince any expert that such sudden movements of huge masses of loess can only be explained by a temporary, very drastic reduction of shear strength of the material, as compared to its original strength before the earthquake.

This raises the question of whether the pores in the lower strata of the loess were saturated to a degree that made liquefaction possible or whether, during the collapse of the loess structure, the stress was temporarily taken up by the large volume of the air enclosed in the pores. Casagrande believes that the liquefaction was caused by the air pressure building up in the voids. That the air enclosed in the pores of powdery material may cause a flow can be observed on a smaller scale when, for example, flour or cement is quickly poured from a bag. However, if huge masses of loess, several hundred meters high, are shaken so strongly that their structures collapse, then it is understandable that it requires a longer period of time until the excess pressure of the air in the pores decreases again. During this period the shear strength of the material is at a very low value. If we can assume the credibility of the report, this may explain the great velocity by which a whole mountainside of loess can cover many square kilometers of a valley "within seconds," burying villages and damming up rivers.

Regarding the predictability of such loess slides, cracks are bound to show at the head of a potential slide or signs of failure should be observable at the foot, but a sudden strong vibration may cause the collapse of the relatively weak loess structure and thereby its liquefaction. As a stabilization measure the soaking of the endangered area with water was suggested, so that the formerly loose soil structure settles under the effect of the water and becomes stable and firm. When carrying out this treatment, care must be taken not to create an excess pore-air pressure; therefore, soaking by sections is recommended.

3.4 Physicochemical Structure Changes of Silt and Clay Soils
F. Hilbert

The most important structural changes in clay minerals are described in detail in 7.2.2.1 through 5. This chapter mainly aims at providing the reader less interested in theory with a summary of the most common causes for landslides in cohesive soils as far as they may be attributed to such changes.

Two decisive factors contribute to a loss of shear strength in clay soils that is sometimes quite considerable: on the one hand, the water absorption and the resulting swelling; on the other hand (and often simultaneously), an ion exchange, whereby ions only loosely bonded in clay minerals are replaced by others. Frequently these two processes interact and may accelerate each other.

Of course, absorption of water, as well as ion exchange, is possible only if water, or in the case of ion exchange, salts dissolved in water, penetrate the soil. In a more empirical sense, therefore, any event that furthers the penetration of water or saline solutions into silty clay or clay soils may be considered a natural cause for a loss of shear strength and eventual landslides. By this definition, there are a great number of causes of landslides, some of them natural, but others quite frequently due to human activities. Although incomplete, the main causes are listed below in order to provide the practicing geotechnical engineer with an overview of the danger of landslides in silty clay and clay soils, and an explanation of how they may be triggered by construction acitivities, irrigation, sewage waters, extreme weather conditions, etc. Special emphasis is placed on the inadvertent triggering of landslides in the course of the construction of transportation routes, water and sewage service lines, and the like. Owing to a lack of knowledge and care, mistakes may be made that could be easily avoided, and that may later cause damages that either require expensive stabilization measures or are irreparable.

3.4.1 Relief from Load Pressure and Resulting Water Absorption

Consolidated clay or silty clay, when relieved of load pressure, regains the ability to absorb water and to swell, whereby the strength decreases proportionally (7.2.2.2). This water absorption as such is a very slow process, because it is difficult for water to penetrate cohesive soils. However, as outlined in detail in 3.3.4, the swelling leads to the development of cracks and fissures that allow water to penetrate farther, and under certain circumstances the softening process may advance rather quickly.

3.4.2 Increase of Water Pressure in Soils

An increase of water pressure in the soil, like the relief of load pressure, may lead to water absorption. The difference between the effective consolidation pressure and the pressure of water in contact with silty clay or clay soils is the decisive factor for water absorption and proportionate loss of strength. Consequently, a consolidated soil may swell if only hydrostatic pressure increases and earth pressure remains constant.

When calculating the stability of a slope susceptible to slides, it is therefore frequently not sufficient to put into the force diagram only the increased pore-water pressure, as in 4.1, Examples 3 and 4; it also needs to be taken into account that at the same time the shear strength of the soil may deteriorate considerably. What part the loss of strength will play in relation to an eventual landslide is a question of the duration of action and the relative height of the pore-water pressure. If the latter rises quickly to the value of earth pressure, then it may trigger the slide by itself, as illustrated by Case History 6.1.4.1 (II). If the increased pore-water pressure alone is not sufficient to endanger the stability of the slope, the water uptake caused by it over a longer period of time can set off a landslide by decreasing the shear strength of the soil.

Since an increase of hydrostatic pressure has, in any case, a harmful effect, every construction activity or natural event that increases this pressure may in turn lead to a landslide. Such pressure increases are—for example—caused by:

3.4.2.1 Increased Water Inflow to Water-Bearing Layers
If the capacity of the natural water outlet is exceeded, the backwater will increase the hydrostatic pressure. Various construction activities, and also natural events, may be the cause for such an increase of water inflow. For example, construction of motorways, construction of railways, grading work, and excavation of construction pits are activities that frequently allow rainwater, often from very large areas, to collect in one place. Activities such as the chaneling of rivers and creeks, regulations, construction of dams for water reservoirs, and the construction of water service lines may allow a very large volume of water to reach areas that previously only had a minor water inflow from rainfall. Deforestation and the creation of arable land may allow the quantity of rainwater seeping into the soil to increase considerably, while evaporation decreases (deforestation has an especially harmful effect if the remaining topsoil is strongly eroded by the flow of surface waters). Finally, extreme rainfall often triggers landslides, sometimes with a considerable delay (this, again, points to the role of a decrease in strength in the soil owing to swelling—see above).

3.4.2.2 Closing of Natural Drainage (Springs, Water Outlets)
Closing or constricting natural drainage, something that may happen purposely or inadvertently during construction work, has the same effect as an increased water inflow.

3.4.3 Development of New Cracks and Fissures or the Opening of Water Passages Previously Closed by Impermeable Soil

New cracks and fissures or opened water passages may, of course, also create an increased hydrostatic pressure or enable water to reach zones more quickly and in larger quantities where it may cause swelling and softening. The main causes are construction activities, erosion by water, and extreme dryness.

Sometimes, for example, a safety factor that is too low is chosen for an embankment constructed with earth fill. This may be the indirect cause for a slide: With a safety factor below 1.5 the embankment will not fail immediately, but if it consists of clayey material, cracks will develop. Water enters the cracks, softens the soil, and after a time the embankment slides down. For the same reason it is difficult to stabilize slides once they are in motion: The slide creates new cracks and fissures by which water can enter; this is a self-perpetuating process. Periods of extreme dryness may also set off this mechanism. Certain soils, when they dry out completely, reduce their volume to such an extent that cracks of considerable depth develop. When rainfall sets in again, water reaches great depths in the soil before the shrinkage cracks close again from water uptake.

3.4.4 Salting of Highways—Ion Exchange

When an exchange of multivalent ions (especially calcium ions) against monovalent ions (e.g., sodium, ammonium, or potassium ions) takes place, the capacity of clay minerals to absorb water increases and their strength decreases (7.2.2.3). This loss of strength may take place in spite of an existing consolidation pressure, and even if the pore-water pressure is very low or zero. Therefore the penetration of salt water or sewage into cohesive soils is especially dangerous. In addition, saline solutions, with their high conductivity, increase the permeability of the soil and consequently penetrate much faster than clear water. This phenomenon is due to the fact that the thickness of the electrochemical double-layer at the inner wall of the capillaries (7.2.2.4.1) decreases at higher conductivity, thereby increasing the cross-sectional area available for an unconstricted passage of hydrostatic flow.

The most common sources of water with such detrimental effect are sewage from stables, leaking cesspools and collecting tanks for feces, seeping and trickling fecal and industrial sewage, leaking channels or gutters carrying sewage, salting of roads, over-use of potash and ammonium fertilizers, and decomposition of rocks and soils rich in natrium and potassium.

3.4.5 Adjacent Reducing and Oxidizing Soil Layers (blue clay/brown clay)—Natural Electroosmosis

As outlined in 7.2.2.5, water inflow at the zone of division between reducing and oxidizing soils, caused by natural electroosmosis, may create an electroosmotic excess pore-water pressure, causing in turn a softening and consequently a loss of strength. By this process sharply defined slip surfaces, sometimes only millimeters thick, may develop, for example, between blue (dark) and brown (light to reddish brown) clay; in these zones the soil is like soft soap. It is obvious that dangerous slides of the upper soil layer may set in when such lubricating planes get big enough. The phenomenon is frequent where oxidizing yellow to brown silty clay rests on top of dark to blue mudstone. The same phenomenon, in principle, may be caused by stray current from d.c. power plants.

3.4.6 Quick Clay

In certain parts of Norway, Sweden, and Canada (Quebec Province) the phenomenon of "quick clay" may be observed (Cabrera and Smalley, 1973; Kenney and Drury, 1973; Rosenqvist, 1977). Quick clays are marine sediments of extreme sensitivity (sensitivity degree up to 40); the clay content -40%, $w_L \cong 26\%$, $w \cong 7\%$. When disturbed, these clays flow like viscous fluid, sometimes over great distances (Janbu, 1977b, 1980).

After Bjerrum (1954), this behavior is due to the fact that after these soil layers had risen from the sea, the salt contained in the clay pores was leached out by fresh water and the soil structure was weakened. Söderblom (1974b) explains the occurrence of quick clay with the presence of organic substances, e.g., fertilizers, sewage, etc. (3.4.4).

Another explanation is offered in 7.2.2.3.3, where it is proposed that the phenomenon is the typical thixotropic behavior of a mass consisting of colloidal particles with weak adhesion. The clay particles were very loosely sedimented to the bottom of the sea and never consolidated, so that there is a relatively large amount of water between them.

It is obvious that such a mass of recent origin, when lifted from the sea without being consolidated by a load pressure, tends to liquefy rather spontaneously (La Rochelle, 1975). In Quebec Province a mass of quick clay 5 to 10 m deep, covering an area 500 m wide and 1,000 m long, flowed into the valley of a tributary of the St. Lawrence River within half an hour (La Rochelle, 1980). Broms (1978) reports on a similar case that he observed on the Gota River in Sweden, where the ramming of piles triggered violent quick-clay slides.

Generally, no stabilization measures have been devised as yet for this type of slide. It can only be recommended that areas prone to quick-clay slides be constantly observed and that any vibrations are avoided. Holm (1961) describes the stabilization of slide-prone quick clay by sand drainage.

3.5 Action of Water in Soil

The action of water in the soil is predominantly responsible for landslides. Water has a variety of effects that will be discussed in the following chapters. Strangely enough, the presence and action of water is frequently overlooked in soil explorations and safety calculations. The momentary absence of water does not preclude that it may appear in a most damaging manner, for example, after heavy precipitation. It often takes quite a number of years until water becomes active, for example, as pore-water pressure builds up behind the slopes of cuts. A lot of experience and the study of relevant literature is needed to predict such occurrences, and to take timely preventive measures.

The actions of water that induce landslides are discussed below and, though incomplete, a number of stabilization examples are presented; these are limited to typical cases.

3.5.1 Action of Pore Water

There are different processes by which pore water may have a damaging effect in the soil (Bauer *et al.*, 1980; Bishop, 1957; Skempton, 1977b).

3.5.1.1 Concentrated Pore-Water Action at Potential or Actual Slip Surfaces or at Walls of Cracks in Clayey-Silty "Homogeneous" Soils

This case is typical for conditions in brown London clay, as described in detail by Skempton (1977a,b). His observations regarding the generation of pore-water pressure at the slip surface in seemingly homogeneous brown clay are very interesting. Pore-water pressure continuously increases over many years until it finally leads to retarded landslides (Skempton, 1977a,b). It has not been established where this water comes from, perhaps by hydrostatic pressure from an uphill water source, and it is questionable whether in such cases the installation of slightly rising drainage pipes would lead to a stabilization.

3.5.1.2 Concentrated Pore-Water Action at the Zone of Division between Clayey-Silty Soils of Different Nature

Pore-water action at the zone of division is quite frequent; the author was able to study such conditions in southern and central Italy, in many parts of Austria, and in Japan (these zones of division where water pressure builds up, discussed in 3.5.1.1 and 3.5.1.2, are quite thin—often only several millimeters thick). Pore-water pressure building up between soil layers with different chemical and physical properties often causes spontaneous landslides. Hydrostatic pressure may build up on account of a connection with an uphill collecting area supplying the water but may also be explained by electroosmotic processes. These phenomena at the zone of division, their effects, and which countermeasures to take are presented in detail by F. Hilbert in chapter 7.

3.5.1.3 Pore Water Acting in Relatively Permeable Silty Sand Layers of Several Decimeters Thickness Interposed between Two Relatively Impermeable Silty-Clayey Layers

The stratified deposits of Sarmat and Pannon, e.g., near Graz, Austria, are frequently composed of a relatively permeable layer of silty sand, interposed between two relatively impermeable silt layers. The layers usually have a slope-parallel dip with gradients of 9° to 11° or less; in many cases the sand layer wedges out at the surface and shows distinct wetness. During periods of heavy rainfall, water flow in the sand layer may become so strong that overpressure builds up, causing the sliding of the upper impermeable layer owing to a decrease of effective stress (4.1). Spontaneous landslides of several meters may take place within a few days (see 6.1.4.1 (II)).

*3.5.1.4   Pore Water Acting More or Less Uniformly in the Whole Mass of
         Clayey-Silty Layers of Several Meters Thickness*

Here the streaming pore water sets the entire slope in motion with relatively
small quantities of water per time unit (no continuous strong water supply).
This case is similar to the one described in 3.5.2.2, but deviates regarding the
water volume per unit time.

In the absence of a continuous strong water supply, slopes of clay, clayey
silt, and sand may be rendered unstable by streaming water (4.1, Example 3).
With a slope-parallel continuous flow, safety η, representing the value for a soil
without water flow, decreases from $\dfrac{tg\varphi'}{tg\beta}$ to $\dfrac{tg\varphi'}{2tg\beta}$ (at $\gamma' = \gamma_w$; $c' = 0$). If a very
strong water source is lacking and the water supply is at times interrupted,
then the slope will temporarily stabilize but starts to slide again as soon as
water action sets in with full force. However, as the grain structure of the soil
prevents the erosion of single grain sizes, the ground becomes a stiff, viscous-
to-pulpy mass that moves only intermittently (contrary to 3.5.2.2), depending
upon the water supply. A trench excavated in this mass will close again within
a short time. Such a soil needs to be drained or it will continue to creep and
endangers any constructions on it or in its vicinity (see 3.5.2.2).

In Case History 6.1.1 (IV), piers were constructed that intersect the sliding
mass so that the soil flows around the piers and the protective constructions,
i.e., the soil continues to move; in Case History 6.1.2 (II), soil movements are
stopped by wells. In Case Histories 6.1.4.2 (I), 6.1.4.3, and 6.2.9.3, the slopes
were drained by slightly rising subsurface drainage lines (in combination with
pile walls), and slides were stopped completely (Holtz and Bomann, 1974;
Holtz and Massarsch, 1976). The choice of the stabilization measures depends
upon the financial means available for a complete stabilization, which must
always be the objective in the vicinity of major constructions.

3.5.2   Action of Water (from a Constant Strong Source) that is Streaming in Soils

3.5.2.1   Sandy Soils

Water streaming into and out of uniform silty sand soils may cause a retrogressive erosion. Case History 6.2.4 describes how strong water flow set sand
in motion so that it was flowing out like a thin liquid. This soil material, also
called quicksand, sometimes only has a gradient of 1° to 2°, and since it develops quite spontaneously, it causes great damage. Quite justly, quicksand is a
menace for the designing engineer.

Stabilization alternatives include:

(1)   Extensive soil exchange, i.e., removal of quicksand.
(2)   Solidification of the quicksand horizon by grouting with chemicals.
(3)   Drainage by slightly rising horizontal drainage lines (Kenney *et al.*,
      1977). Usually it is also necessary to install a Terzaghi filter where the
      sand layer wedges out at the slope toe.

In Case History 6.1.4.2 (II) the outflow of sand caused a whole industrial plant to sink 20 m. Here it would have been necessary to install slightly rising drainage pipes and to protect the toe of the slope with a Terzaghi filter.

3.5.2.2 Clayey-Sandy Silt Soils

Water flowing in and out of a slope of clayey-sandy silt will only cause movements if a constant strong water source is present. Under such conditions, the fine particles are not washed out nor does quicksand form, but the whole viscous mass creeps downhill. Contrary to the mass described in 3.5.1.4, this mass moves continuously (e.g. several millimeters per day), passing without stopping over any constructions, roads, channels, etc. This type of movement of an embankment toe will cause the embankment to cave in. Chapter 6.2.3 presents the case history of a stabilization where water flowing from permeable limestone mountains dislocated the embankment toe until the roadway failed. The pulpy soil was replaced by a stone wedge which provided the necessary drainage and also increased friction strength in this critical zone.

Case History 6.1.4.1 (II) presents another example for the stabilization of constantly and heavily watered clayey-sandy silt layers. It was decided to excavate up to 5-m-deep drainage trenches, at the bottom of which plastic tubes were placed in gravel and wrapped in filter fabric. The installation of slightly rising, bored drainage pipes was not considered because the affected area was too large for this method.

In summary, the following may be said regarding slide stabilization by drainage: The basic feasibility of stabilizing active landslides by drainages is undisputed; it has been carried out with full success at numerous landslides.

The design of the slope drainage requires exact and sometimes very extensive soil-mechanical investigations, and the analysis of hydraulic conditions. The former must be carried out to determine the causes of the landslide beyond a doubt; only if ground-water flow is involved should subsurface drainage be taken into consideration. Hydraulic conditions are of decisive significance for the determination of the required depths and the arrangement of the subsurface drainage lines. If preconditions for a slide stabilization by drainage are met, it presents no problem to carry them out properly and at justifiable cost.

Proper stabilization work calls for drainage that will not clog and will have lasting effectiveness. Following this principle and in view of today's technology, it is high time to stop using clay and concrete pipes for subsurface drainage, even though they may be easier to install, since experience has proved that their effectiveness is very limited and short-lived. If clay or concrete pipes are placed in slide-prone subsoil, where slope movements have not yet fully subsided, they may be rendered ineffective by dislocation immediately after installation. Contrary to the original intention, this may even lead to a weakening of stability conditions, caused by water concentrations at the points where the drainage ducts are broken.

Drainage with gravel backfill must be protected against clogging by the fine particles in sandy soils to ensure permanent effectiveness. This may be achieved, for example, by the use of filter fabric. It is also expedient to divert the ground water collected in the trench with plastic pipes. Excavation methods have shifted almost completely from manual labor to machinery, but the principle remains valid that water-bearing layers should be followed as far as possible by the drainage system. Nowadays the geotechnical engineer has a wide range of machinery at his disposal and with some experience it is even possible to use machinery with good results on marshy ground with virtually no bearing capacity. Operating depths of 5 to 6 m are absolutely feasible; greater depths require preliminary excavations. Various regulations for the protection of the worker demand that drainage trenches be braced, but where the soil is sufficiently cohesive, the sides may be left free-standing. Of course, excavation work can be carried out in longer sections if the trenches are braced.

The main advantage of mechanical excavation is its speed. Cohesion, which is present in most soils, remains effective at least over a short period of time, and this has a positive effect on the stability of the slope. Reduction of pore-water pressure due to soil expansion during excavation work has an analogous effect.

Water flow in drainage in relation to time decreases sharply as compared to the initial discharge; after some time the quantity of water collected may amount to only a few percent of the initial volume. This indicates not only the effectiveness of the drainage but also the speed of soil consolidation in the area of the original slide.

Lastly, it should be pointed out that the design of the drainage, even if based on the most exact soil investigations, should not represent an irreversible construction plan. Optimal stability is only achieved by deciding in the course of construction work on the final positions and depths of the drainage lines. Of course this requires great experience and know-how on the part of the contractor, and a great deal of trust between client and contractor.

3.5.3 Action of Flowing Surface Waters

3.5.3.1 Water in Vertical or Steep, Fissured Soil Layers
As an exception, a slide of rocklike subsoil—sandstone with a 45° dip of the layers, vertical fissures, and frequent interposed marl layers—and its stabilization are described here. Surface water entered the fissures, distended them, softened the marl, and reduced its shear strength. Owing to the slope-parallel stratification, whole blocks slid down and landed on the railway below (see Case History 6.1.4.1 (I)). Here the expensive construction of a retaining wall could be avoided by a number of measures, especially the collection of surface waters in open drainage channels.

3.5.3.2 Water Mixed with Superficial Soil (Debris Avalanches)
Water flowing on the surface, together with the action of weathering, loosens the structure of the topsoil. When the flow of water increases, for example during heavy rainfalls, a mixture of water, humus, and fractured and weathered rock and soil rushes downhill as a pulpy mass, destroying and carrying

along trees, roads, and buildings. These are the much feared debris avalanches, which can only be prevented by extensive constructions; but this is a matter of "mountain torrent regulations" and will not be treated here.

3.5.3.3 Water Undercutting Slope Toes

As described in Case History 6.1.1 (I), flowing water, for example, a river, may cause the toe of a slope to slide. To achieve a permanent stabilization, additional measures are of course required, such as the construction of a wall following the course of the riverbank, to protect the slope toe against further erosion and to prevent undercutting.

3.5.4 Solifluction

Solifluction is a special type of sliding and flowing strata close to the surface, which today may only be observed in the Arctic and in high mountain regions. There the soil thaws only to shallow depths during the short summer seasons. Water from melting and precipitation saturates the topsoil and it flows like heavy snow slush, even on rather level slopes (2° to 4°); the subsoil is impermeable since it is firmly frozen. During the Pleistocene, solifluction was quite common and an important factor in the formation of the terrain morphology.

Today one finds larger and smaller lobate pieces of rock firmly embedded in a matrix of sand, silt, and clay that is 3 to 5 m thick and often very hard. The subjacent clay contains slickensides even if the ground surface has only a 3°-to-4° gradient (Skempton and Hutchison, 1969). These fragments are frequently covered by loess or residual clay and the areas carry a lush vegetation. It may be assumed that they originate in the Earth's history and that their formation took place in the ice age under conditions similar to those in polar regions today.

Biczysko and Starewsky (1977) describe a landslide that took place during the construction of a bypass for Daventry. A layer of blue clay from the Upper Lias, subjacent to brown silty clay and clayey sand (Northampton Sand Ironstone) was exposed by the cut. During the last ice age, this soil had undergone alternating deep freezes and complete thaws, and had moved downhill in numerous small landslides. The zone of division only had a 9° gradient, which decreased to 4° near the valley floor.

The upper layer was about 2.5 m thick, showing on its surface no signs of previous slide movements when construction work for the road was started. At a depth of approximately 3 m, that is 0.5 m below the zone of division, strong slide movements set in during excavation work—obviously taking place on an old slip surface stemming from "solifluction"—thereby setting the upper layer in motion.

The angle of friction of residual strength was 13°; pore-water pressure at a depth of 2.8 m below the surface was 0.15 bar when the first slide movement occurred, decreased during the "breaking" of the upper layer to 0.075 bar, and finally, during complete "disintegration" to 0.04 bar. Safety calculations after

Bishop yielded a safety factor of 1. Biezysko and Starewsky describe the difficulties encountered when carrying out cuts in such soils, and show that the relocation of the road alignment to an uphill zone was the only possible alternative.

The following factors make it especially difficult to assess the danger of landslides:

(1)   The surface of these slopes, with gradients of 10° or less, showed no signs of previous slide movements.

(2)   Old shear planes were not discovered by exploratory borings in the critical soil, as they were obliterated by the extracting tools and no longer recognizable.

(3)   Shear planes were only recognizable in samples that were extracted at a 45° angle of axis to slope, and subjected in the laboratory to cylinder compression tests.

(4)   Reliable results regarding shear strength could be obtained in the triaxial apparatus, provided the plane of failure of the test cylinder was inclined by approximately 50° to horizontal.

(5)   The friction angle amounted to only 13°.

(6)   The height of the ground-water table and its changes has to be registered precisely by piezometers.

Therefore, it may be stated in summary that the surficial crust of, for the most part, only slightly inclined solifluction layer had settled after the actions of freezing and thawing ceased, and did not show signs of previous slide movements. The subjacent, blue clay layer, however, was greatly disturbed and weakened in places by slide movements and the resulting shear cracks (shear angle approximately 13°).

If such blocks of layers are disturbed in their equilibrium by a cut at the toe, then spontaneous slides will set in, whereby the height of the ground-water table may play an important part.

4 Theoretical Basis for the Calculation of Slope Safety

As outlined in 1.1, there are an extraordinary variety of effects that may cause a landslide, so preconditions for the stability of a slope escape strictly theoretical calculations. Calculations of slope safety will meet the objective only if exact data on marginal conditions and accurate soil-mechanical analyses are available. Of course, one must also remain aware of the fact that small irregularities and inhomogeneities in the soil, which were not discovered during soil exploration, may render the results of the calculation completely invalid. Such factors of disturbance are, for example, thin, water-bearing layers, slip surfaces of previous slides, and fine fissures and cracks with different courses. Nonetheless, some basic calculation methods for the determination of safety (Colleselli, 1977; Soos, 1970) are presented below.

4.1 Safety Conditions in a Slope with Slope-Parallel, Planar, Infinite Slip Surface

Examples for calculating the safety of a slope above a slope-parallel, planar, infinitely long slip surface are given for the following conditions:

(1) Cohesionless soil, no ground water, no pore-water pressure
(2) As in 1, but with earthquake action
(3) As in 1, but with slope-parallel ground-water flow above the slip surface
(4) As in 1, but with pore-water pressure
(5) As in 1, but with cohesion

Examples 1 through 4 correspond to conditions as they prevail in nature in homogeneous soils and in cohesionless soils that are stratified parallel to the slope surface. Example 5 is only relevant for cohesive soils with a slope-parallel stratification. If example 5 is used with homogeneous soils, then a circular slip surface—as it occurs in nature—must be assumed for the calculation of limit equilibrium.

Definition of symbols used in figures and equations:

h = height of slide layer
l = length of examined soil prism
b = breadth of examined soil prism; $b = 1$
d = distance between ground surface and slip surface
β = angle of slope to horizontal $\equiv$ gradient of slip surface
γ = unit weight of moist soil (kN/m³)
γ' = unit weight of submerged soil (kN/m³)
γ_r = unit weight of water-saturated soil (kN/m³)
γ_w = unit weight of water (kN/m³)
φ' = effective angle of internal friction
c' = effective cohesion intercept
σ = stress

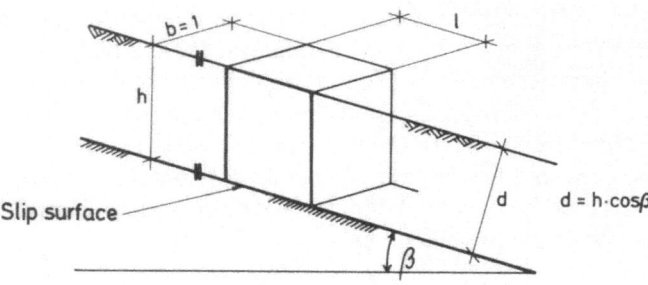

Figure 4-1   Slice of planar slide layer.

1   Cohesionless Soil, No Ground Water, No Pore-Water Pressure (Fig. 4-2)

The soil may be moist (γ), dry (γ_d) or water-saturated (γ_r), it may also be under uplift (γ'), i.e., a submerged slope.

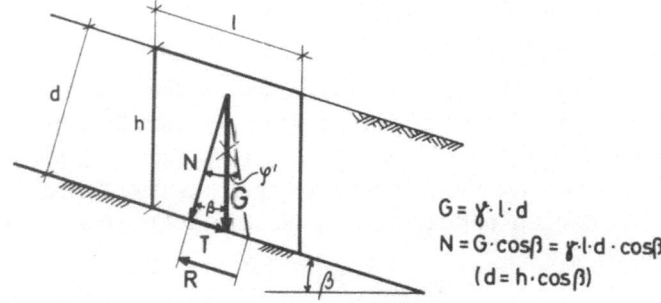

$$G = \gamma \cdot l \cdot d$$
$$N = G \cdot \cos\beta = \gamma \cdot l \cdot d \cdot \cos\beta$$
$$(d = h \cdot \cos\beta)$$

Figure 4-2   Gravitational forces in a planar slide layer.

Driving force: $\quad T = N \cdot \mathrm{tg}\,\beta \equiv G \cdot \sin\beta = \gamma \cdot l \cdot d \cdot \sin\beta$

Resisting force: $\quad R = N \cdot \mathrm{tg}\,\varphi' = \gamma \cdot l \cdot d \cdot \cos\beta \cdot \mathrm{tg}\,\varphi'$

Safety factor: $\quad \eta = \dfrac{R}{T} = \dfrac{N \cdot \mathrm{tg}\,\varphi'}{N \cdot \mathrm{tg}\,\beta} = \dfrac{\mathrm{tg}\,\varphi'}{\mathrm{tg}\,\beta}$

2   As in 1, But with Earthquake Action (Fig. 4-3)

The force of earthquakes H_E is stated as a percentage of the total weight of the soil and water, i.e., it is not calculated with γ'!

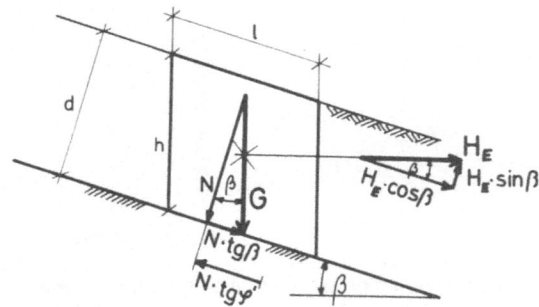

Figure 4-3   Forces in a planar slide layer from gravity and earthquake action.

Driving force: $\quad T = N \cdot \mathrm{tg}\,\beta + H_E \cdot \cos\beta$

Resisting force: $\quad R = (N - H_E \cdot \sin\beta) \cdot \mathrm{tg}\,\varphi'$

Safety factor: $\quad \eta = \dfrac{(N - H_E \cdot \sin\beta) \cdot \mathrm{tg}\,\varphi'}{N \cdot \mathrm{tg}\,\beta + H_E \cdot \cos\beta}$

3   As in 1, but with Slope-Parallel Ground-Water Flow Above the Slip Surface (Fig. 4-4)

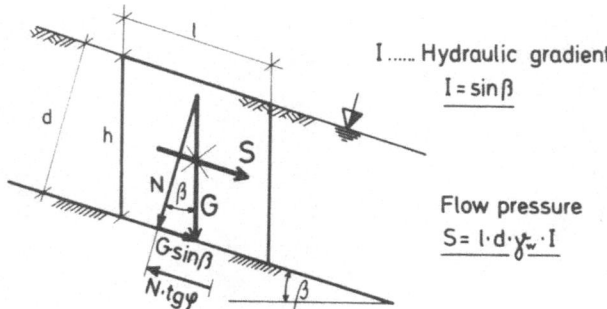

I ...... Hydraulic gradient

$I = \sin\beta$

Flow pressure

$S = l \cdot d \cdot \gamma_w \cdot I$

Figure 4-4   Forces in a planar slide layer from gravity and slope-parallel ground-water flow.

Driving force: $T = G \cdot \sin \beta + S = l \cdot d \cdot (\gamma' \cdot \sin \beta + \gamma_w \cdot I) = l \cdot d \cdot \sin \beta \cdot (\gamma' + \gamma_w)$

Resisting force: $R = N \cdot tg\, \varphi' = G \cdot \cos \beta \cdot tg\, \varphi' = l \cdot d \cdot \gamma' \cdot \cos \beta \cdot tg\, \varphi'$

Safety factor: $\eta = \dfrac{l \cdot d \cdot \gamma' \cdot \cos \beta \cdot tg\, \varphi'}{l \cdot d \cdot \sin \beta\, (\gamma' + \gamma_w)} = \dfrac{\gamma' \cdot tg\, \varphi'}{(\gamma' + \gamma_w) \cdot tg\, \beta}$

with $\gamma' \simeq \gamma_w \rightarrow \eta = \dfrac{tg\, \varphi'}{2 \cdot tg\, \beta}$

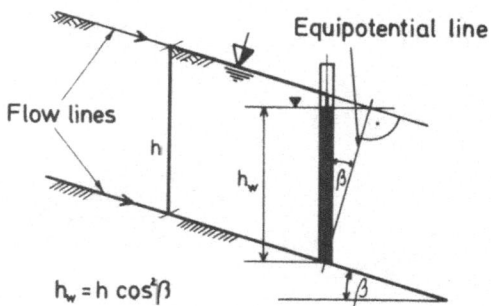

Figure 4-5   Value of pore-water pressure from slope-parallel ground-water flow.

One arrives at the same safety factor by defining $\gamma_w \cdot h_w$ as pore-water pressure, as in Fig. 4-5, and then calculating safety as in example 4 by using γ' instead of γ_r.

4   As in 1, but with Pore-Water Pressure (Fig. 4-6)

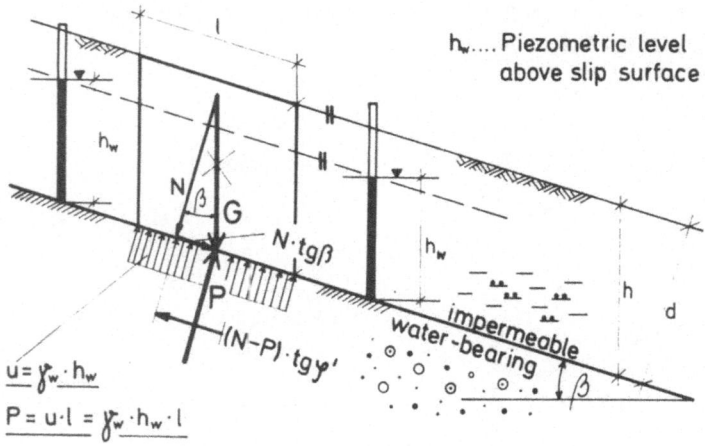

Figure 4-6   Forces in a planar slide layer from gravity and pore-water pressure.

Driving force: $T = N \cdot \text{tg } \beta$ $\qquad$ G is calculated with γ or γ_r

Resisting force: $R = (N - P) \cdot \text{tg } \varphi'$ $\qquad$ (never with γ'!)

Safety factor: $\eta = \dfrac{(N - P) \cdot \text{tg } \varphi'}{N \cdot \text{tg } \beta}$

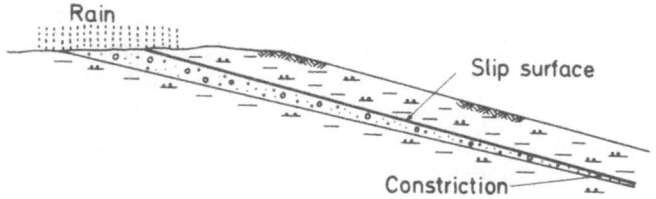

Figure 4-7 Pore-water pressure due to constriction of the permeable layer.

A quite thin, water-bearing layer may exert pore-water pressure on the superimposed, virtually impermeable slide layer, if the water is banked up, e.g., during or after heavy rainfalls (Fig. 4-7). Pore-water pressure may also be caused by quick loading (filling operations).

5 As in 1, but with Cohesion c' (Fig. 4-8)

In cohesive soils, as a rule, circular rather than planar slip surfaces will develop. In cohesive soil a planar slip surface will only occur if preconditioned by a stratification.

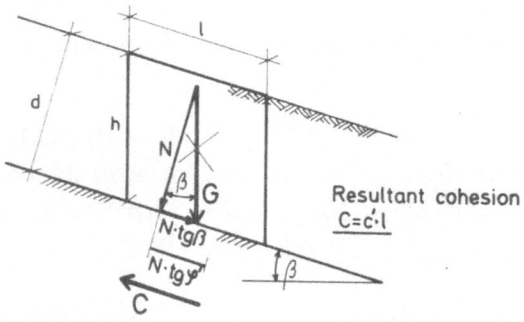

Figure 4-8 Forces in a planar slide layer from gravity and cohesion.

Driving force: $T = N \cdot \text{tg } \beta$

Resisting force: $R = N \cdot \text{tg } \varphi' + C$

Safety factor: $\eta = \dfrac{N \cdot \text{tg } \varphi' + C}{N \cdot \text{tg } \beta}$

As shown above, reduction of cohesion with time (e.g., soaking or decaying diagenetic cohesion) leads to reduction of safety.

Summary

Various combinations of examples 1 through 5 may be encountered. A change of weight as such, due to rainfall (water absorption) will not of itself cause a change in safety conditions (see example 1). But rain will affect the safety of a slide layer if it causes an increase of pore-water pressure or if soaking decreases a present cohesion or reduces the angle of friction (shear strength) by the formation of a lubricating layer.

Finally, the possibility of crack-water pressure must also be taken into account (Fig. 4-9). In the zone of an incipient landslide cracks will develop; these may also be shrinkage cracks caused by a drying-out of the soil. During rainfall the cracks fill with water and therefore the pressure of crack-water must be taken into account as an additional driving force in the analysis of safety in a slide layer.

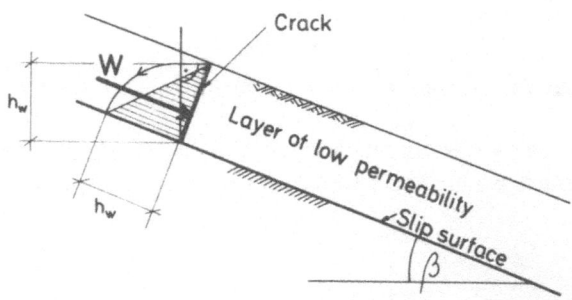

Figure 4-9   Action of crack-water pressure upon safety in a planar slide layer.

4.2   Methods for Calculating Slope Safety Regardless of Subsoil Structure *(Synopsis of Different Methods to Determine Safety of Slopes with Slip Surfaces of Any Given Shape (after Eigenberger, 1972))*

Though in cohesionless soils without water flow, independent of the thickness of the layer, the expression $\eta = \mathrm{tg}\,\varphi'/\mathrm{tg}\,\beta$ is valid for safety (after Fellenius's rule $\eta = \mathrm{tg}\,\varphi'_v/\mathrm{tg}\,\varphi'_e$; φ'_v is the present angle of friction, φ'_e the required angle of friction), conditions in commonly encountered soils are more complex. It is known from experience that landslides in cohesive soils have more or less circular–cylindrical slip surfaces. This shape is the basis for a multitude of methods for the determination of slope safety that the geotechnical engineer has at his disposal. The safety factor η, arrived at by the different methods, often varies considerably (Smoltczyk, 1975).

When safety has to be determined, the following questions come up:

(a)   How to define safety?
(b)   How to determine the position and shape of the slip surface?
(c)   Which calculation methods yield the most accurate results regarding safety?

4.2.A   Definition of Safety

Though peak strength is governed by a modified elasticity theory, *a modified plasticity theory is valid for safety,* which yields not only *simpler* but also better results. Therefore the definition of safety should meet the following criteria:

(a)   The intersecting forces causing failure are relevant.
(b)   Since the plasticity theory is valid, the intersecting forces must be independent of the path of displacement. Coulomb's law is valid.
(c)   Uncertainty of strength is of much greater significance than uncertainty of forces.
(d)   In the calculation of total safety the "characteristic values" of strength must be determined.
(e)   The probability of the combination of forces must be observed.
(f)   Fluctuations in the initial values are reflected best by partial safety coefficients.
(g)   Values used in the calculation should be as constant and as accurate as possible.
(h)   The definition should be applicable for any given shape of test surfaces.
(i)   The definition should be applicable for any soil coefficients and layers.

The following existing safety definitions were examined.

(a₁)   State of Failure Due to Application of External Forces:
Hultin and Petterson Definition
Fellenius's Definition *for friction soils*
Borowicka's Earth Pressure Rule
Fröhlich's Definitions

These definitions mainly violate (c) and in part (g), (h), and (i). From this group Fröhlich's Definition No. 2 is the best.

(b₁)   State of Failure Due to Decay of Soil Coefficients:

(α). Comparison between present and required shear strength:
Fellenius's Rule
Sior's Definition
Definition after Terzaghi and Peck (1961)
Ohde's Rule
Lazard's Rule

These definitions correspond better to actual conditions than those in group (a_1).

(β). Reciprocal degrees of mobilization as definition of safety:
These definitions were published by Taylor. Only part of the effective forces is used for the calculation of safety against failure and this represents a violation of (a). Furthermore, (i) is not observed (exception—"true safety").

(γ). Forces arranged in the direction of rotation (Terzaghi and Peck, 1961). The moment of available as well as required shear strength is falsified by a value $M_2 = G_2 \cdot l_2$. Also, (h) is violated.

(c₁)   Partial Safety Coefficients
Their application is acceptable; however, the same coefficients may not be used for all soils. Disadvantage: Total safety cannot be determined.
   Three definitions of safety correspond well to criteria (a) through (i), namely:

> Partial safety coefficients (for disadvantage, see above), Lazard's Rule (only for homogeneous soils) Fellenius's Rule (advantage—may also be used for stratified soils, yields total safety; therefore the soundest definition).

4.2.B   Position and Shape of Slip Surfaces

One needs to differentiate between linear and planar failure. In almost all methods for determining slope safety, *linear failure* is assumed.
   Theoretical analyses yield the following:

Semi-infinite body under dead load:
   Friction soil—planar slip surface
   Cohesive soil—planar slip surface if slope angle to horizontal $\beta = 0$; otherwise curved.
Break in terrain: Three zones, the surface zones as in the semi-infinite body, interposed curved surfaces, no break.
Linear/planar failure: Slip surfaces have different shapes.
Gravitational forces change direction: Curved slip surfaces.

Observations, Experiments, Comparative Calculations
Linear failure predominates by far.

Homogeneous Soils:
   Friction soil: level, approximately planar slides.
   Cohesive soil: curved slip surfaces, reaching deeper the larger the cohesion component.
   Shape: circles, cycloids.

Stratified Soils:
Layers with great strength are avoided or intersected by the path of least resistance. This refers to steep slip surfaces near the slide head, provided the layers there are sound.

Comparative calculations show that it is sufficiently accurate to assume planar test surfaces in friction soils and circular-cylindrical slip surfaces in cohesive soils. In stratified soils the accuracy of these approximations depends upon the course of the stratification.

4.2.C   Which Calculation Method Yields the Most Accurate Results Regarding Safety?

(A₁)   Planar Test Surfaces
Here all methods based on Fellenius's Rule yield the same values, since stress distribution is irrelevant for the determination of safety.

(B₁)   Methods Based on Kötter's Equation
To these belong

Frontard's Method
Jaky's Method
Investigations by Rodriguez
Brinch Hansen's Equilibrium Method
Characteristica-Method

Stress distribution along slip surfaces is determined after Kötter. In the first four methods the shape of the slip surface is selected; this actually renders Kötter's Equation invalid. The constants of the equation are determined from marginal conditions. The first three methods result in a violation of equilibrium; as a result the calculated value η is too small. Brinch Hansen's Method only violates the "physical possibility" of stress distribution and probably yields better results. The Characteristica-Method assumes a planar failure. If the actual surface corresponds to that arrived at by calculation, then the result is accurate. This group of methods only has a limited applicability, either because of the work involved or because of preconditions.

(C₁)   Methods with Curved Test Surfaces

(α₁). "Exact" methods
To these belong

Fellenius's Graphical Method
Method of Logarithmic Spirals
Morgenstern and Price Method

"Exact" has been set between quotation marks because these methods also only yield results with certain fluctuations ($\Delta\eta \cong 2\%$), owing to the free play of physically possible stress distributions. The first and the third method are superior to the second because they allow variations of the test surface shape (the third method is the computerized mathematical analysis of Fellenius's Method).

(β$_1$). Approximation methods
A stress distribution that is not quite exact is assumed. One or several of the following criteria affect the accuracy of results:

- Shape coefficient $\zeta = \Sigma\sigma N'/N'$: this is very constant
- Absolute value of N: this is mainly a function of direction and magnitude of divergence
- Direction of N': this is rendered erroneous by a unilateral shift of stress distribution
- Position of reference point $\Sigma M = 0$: valid for any test surface shape.

The great number of methods published may be categorized in three groups:

(1) All equilibrium conditions of the total system are met. The physical possibility of stress distribution is not checked.
(2) Safety η is calculated from $\Sigma M = 0$. The force of support regarding its position and magnitude approximately equals the resultant (divergence).
(3) Safety η is calculated after the force diagram. Also, moment conditions are not checked here. Regarding an erroneous, unilaterally displaced σ'_n-distribution, the methods in this group are more sensitive than those of group (2).

Re (1)—to these belong:

Hultin-Petterson's Method, using the Fellenius Rule
Various friction circle methods
Bell's Method
Bishop's Improved Method
Nonveiller's Improved Calculation Method
Spencer's Method

The improved friction circle methods contain somewhat fewer errors than the other methods, since errors caused by an "unreasonable stress distribution" are eliminated.

Re (2)—in this group are the most commonly used methods, namely:

Krey's Simplified Method
Bishop's Simplified Method
Nonveiller's Simplified Method
Franke's Method
Berth's Method
Fellenius's Approximation
Terzaghi's Method (ordinary method)

Here results mainly depend upon how far the *absolute value of normal strength* corresponds to the "true" value. It is believed that *Bishop's* improved method is the best because *a very favorable direction of divergence has been selected*. It has virtually the same accuracy as the methods in group (1) and—with the exception of the friction circle methods—involves considerably less work.

Re (3)—to these belong:

- Method of the German Experimental Institute for Water and Naval Engineering, Berlin
- Janbu's Simplified Method
- Janbu's Improved Method
- Borowicka's Calculation Method

In order to obtain accurate results with these methods, the gradient of normal force must correspond to the "true" gradient. Unfortunately this value is rather error-prone. Here Borowicka's method yields the best results.

4.3 Critical Evaluation of Methods of Calculating Slope Safety and Suggestions for Improvement (Eigenberger, 1972)

4.3.1 Remarks in Principle Regarding the Methods Listed in 4.2.C

The following general presumptions apply to all methods for the determination of slope safety:

(a) Soils are, at least in some layers, homogeneous and isotropic.
(b) Coulomb's Law is valid for failure conditions $\tau_r = c'_r + \sigma'_N \cdot \tan \varphi'_r$
(c) The state of stress is planar.
(d) All acting forces are known.
(e) The law of effective and total stress is valid: $\sigma' = \sigma - u$

The following three conditions should be met by the various methods:

(a) Coulomb's Failure Law must be valid at least along one continuous test surface.
(b) In accordance with the safety definitions of Fellenius, failure along the test surface should occur by a decay of soil coefficients while the degree of mobilization remains constant ($\kappa_c = \kappa_\varphi$).
(c) The forces acting upon the test mass should be balanced.

All three conditions are met if an exact solution of Kötter's equations (differentiated by linear and planar failure) is obtained, given an equal degree of mobilization of cohesion and friction as well as a safety factor $\eta = 1$. Kötter's equations indicate the change of stress along a curved slip surface. So far, a closed solution of Kötter's equations (a partial differential equation) has been obtained only for special problems; in the common case only approximation methods can be used.

4.3.1.1 Planar Test Surfaces

Exact solutions are obtained for planar, infinitely long, homogeneous, cohesionless slopes. There is an equilibrium of forces. All normal stresses σ'_N act parallel to each other; therefore direction and value of normal strength N'_e may be clearly determined. Planes with a parallel course have the same safety coefficient, whereby those parallel to ground surface have the lowest values (Fig. 4-10).

If pore-water pressure acts upon the test surface, then the specific-gravity-load case is valid, as shown in Fig. 4-10a:

$$\eta = \frac{(G \cdot \cos \beta - u \cdot \Delta l) \tan \varphi'_v}{G \cdot \sin \beta}$$

With transverse gravitational forces the expression is analogous.

In the exceptional case of a finite slope, ground surface represents the most critical test surface, unless external forces act upon the slope. Therefore the same conclusions are valid for these surfaces as for the infinite slope.

However, if external forces affect the slope, as in Fig. 4-11, then the assumption of a planar test surface becomes only an approximation; but usually this is sufficiently accurate. Also, in this case normal strength N'_e may be determined accurately, even though exact stress distribution is not known.

Another exact solution is possible for cohesive soils (Fig. 4-12) if a stratification enforces a slip surface. But in this case the distribution of normal stress along the slip surface may not be determined exactly; however, this is of no consequence for safety.

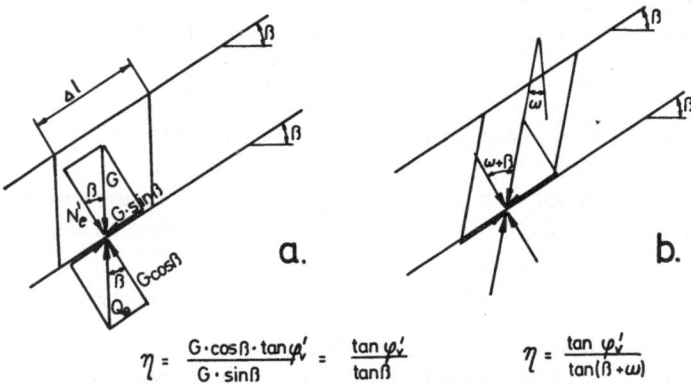

$$\eta = \frac{G \cdot \cos \beta \cdot \tan \varphi'_v}{G \cdot \sin \beta} = \frac{\tan \varphi'_v}{\tan \beta} \qquad\qquad \eta = \frac{\tan \varphi'_v}{\tan(\beta + \omega)}$$

Figure 4-10 Infinite slope of cohesionless soil; a—specific gravity; b—transverse gravitational force.

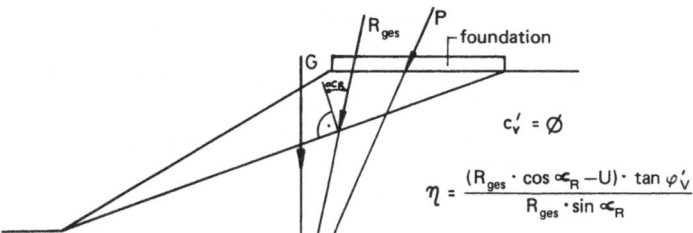

Figure 4-11   Finite slope with $c_v' = 0$ and external forces; U is the resultant of eventual pore-water pressures.

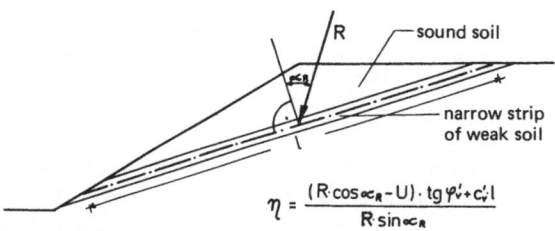

Figure 4-12   Finite slope with $c_v' \neq 0$ and external forces; U is the resultant of eventual pore-water pressures.

4.3.1.2   Curved Test Surfaces

Generally, circular-cylindrical test surfaces are used, but cylinders with logarithmic spirals as directrixes, a combination of planar slip surfaces, or a combination of all three shapes may also be used. If circular cylinders are assumed, this greatly simplifies the calculation. Curved slip surfaces are encountered wherever cohesive forces are active in the soil.

The line integral $\int_0^l c_v' \, dl$ along the test surface can be solved exactly and thereby also the position of cohesive strength. The multitude of methods results from the fact that with curved slip surfaces the system is statically undefined even in the state of failure; therefore, an approximate solution may take many different courses. But only friction strength is statically undetermined, not cohesive strength (Fig. 4-13); therefore—given an equal probability of the soil coefficients—calculation accuracy increases the higher the cohesion component.

In his thesis, Eigenberger first reviews the different trends in the field of calculating slope safety and then analyzes the various methods from the viewpoint of their susceptibility to error, their versatility in application, and the calculating work involved, giving special attention to conclusive reasoning and

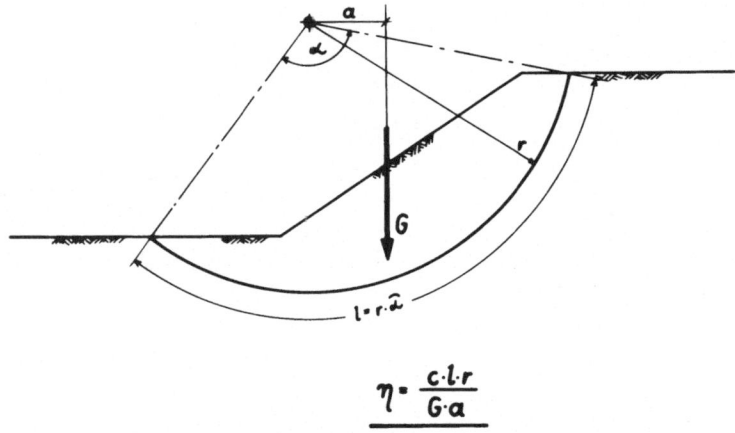

$$\eta = \frac{c \cdot l \cdot r}{G \cdot a}$$

Figure 4-13   Circular-cylindrical slip surface with cohesion only ($\varphi' = 0$).

completeness of references. This analysis reveals strong deviations in accuracy between the individual methods, depending on whether the assumptions are based on constant or strongly fluctuating values (see $C_1 \beta_1$). Among the methods requiring less calculating work, the improved friction circle method and the simplified Bishop method are to be noted for being especially exact.

The improved friction-circle method by Krey (1926) for cohesionless soils, generalized by Taylor (1937) for all soils and load distributions, was further developed by Borowicka (1968) by the introduction of internal pressure in place of cohesion; this represented a very valuable and sensible simplification.

Fröhlich (1949) also developed a simplification, however, he abandoned Fellenius's rule as a safety definition, thus emphasizing the small fluctuations of the shape coefficient. The basis for the accuracy of the improved friction-circle method is that only one assumption regarding the fairly constant shape coeffiicent needs to be made, and that normal strength is determined by a force diagram without divergence. It has the disadvantage that it must be adapted if used with stratified soils and any slip surface shape.

The simplified Bishop method represents one of the best-known lamellae methods. It can be used easily for stratified soils, but involves somewhat more work and will produce larger errors if, by mistake, peak stresses occur at the foot of the test body. The very good results, on the whole, are based on the use of moment condition for the determination of safety and the usually small error in the absolute value of normal stress, due to a favorable (horizontal) direction of divergence.

Eigenberger's approximation method, described below, is based on these findings and represents a further development of the improved friction-circle method. In spite of the small calculating effort involved, it produces results almost as good as the very time-consuming graphical method of Fellenius, which, however, has so far not been surpassed as far as accuracy is concerned.

4.3.2   Simplified Eigenberger Method

This method can be used for homogeneous as well as arbitrarily stratified soils with different load distributions (distributed or single loads) and eventual action of water pressure, and any shape of slip surface (circular-cylindrical, planar, etc., and combinations of these shapes). The most critical slip surface, determined by an approximately correct (i.e., "simplified") force diagram, may then be checked by an "improved" method. However, as may be seen from the examples in 4.3.4, the safety coefficients obtained by the "improved" methods deviate only within a very narrow range from those of the "simplified" method. Therefore only the simplified method is presented below; for the improved method, refer to Eigenberger (1972).

The basic idea behind Eigenberger's simplified method is to determine the approximate absolute value for the normal strength of the total system $|N'_I|$ by way of an approximately accurate force diagram (Fig. 4-14) and to draw conclusions from the shape coefficient $\zeta_{ges} = \dfrac{1}{2}(1 + \zeta_c)$ on total normal stress $\Sigma|\sigma'_N|$. The direction of normal strength is approximated by the straight line $\overline{IM}$, whereby I is the intersection between the resultants of external forces and the slip surface, and M is the center of rotation. As shown by the force diagram of Fig. 4-14, the values for true normal strength $|N'|$ and approximated strength $|N'_I|$ differ very little. The shape coefficient ζ_{ges} is mainly a function of the central angle α_z, and changes insignificantly even if this angle is varied over a wider range ($\alpha_z = 40°$ to $150°$), i.e., ($\zeta_{ges} = 1.01$ to 1.178). Therefore quite exact conclusions may be drawn as to total normal stress $\Sigma|\sigma'_N|$ along the slip surface. The method is illustrated best by the first example of a homogeneous slope with cohesion and friction.

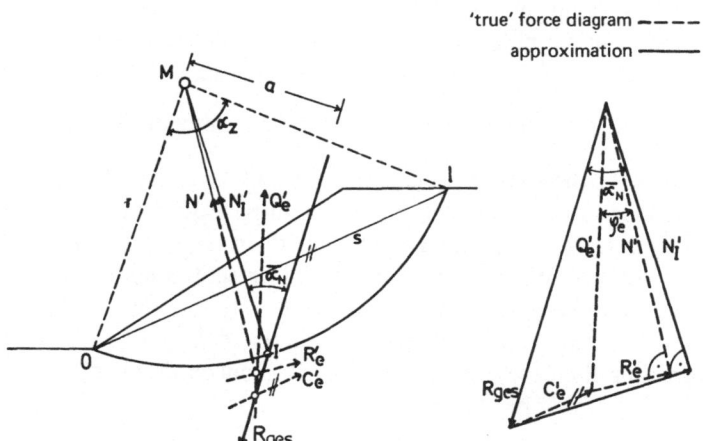

Figure 4-14   Homogeneous soil; approximation after Eigenberger.

4.3.2.1   Homogeneous Soils (Fig. 4-14)

Definition of symbols:

R_{ges} = resultant of external forces (dead load, distributed load, water pressure)

C'_e, Q'_e, N', R'_e = forces of the "true" force diagram. If "true" safety η_w is known, which can only be determined by iteration, even for a given slide circle, then $C'_e = c'_v \cdot s/\eta_w$, out of force diagram Q'_e, and $\tan \varphi'_e = \tan \varphi'_v /\eta_w$, out of force diagram R'_e and N', is valid.

φ'_e = required angle of friction of the true force diagram

N'_I = normal strength of the approximated force diagram

$\Sigma|\sigma'_N| = \int_0^l \sigma'_N \, dl$ = sum of normal stresses

I = intersection of resultant and test circle

l = arc length of test circle in soil

$\overline{\alpha}_N$ = angle between R_{ges} and N'_I

a = normal distance between force R_{ges} and M

r = radius of slide or test circle (slide circle = most critical test circle)

s = cord length

α_z = center angle of test circle

The following relations are valid:

$$\sin \overline{\alpha}_N = \frac{a}{r} \qquad N'_I = R_{ges} \cdot \cos \overline{\alpha}_N$$

The sum of normal stresses $\Sigma|\sigma'_N|$ along the test circle is determined by way of the shape coefficient ζ_{ges}: $\Sigma|\sigma'_N| = \zeta_{ges} \cdot N'_I$.

When using one of the various lamellae methods, however, normal stress due to external forces must be calculated individually for every single lamella and then added up along the test circle.

The significant simplification arrived at by Eigenberger lies in just this difference in the calculation of total normal stress, i.e., $\Sigma|\sigma'_N|$ after Eigenberger versus $\int_0^l \sigma'_N dl$ after the various other lamellae methods. The latter yield different values for the sum of normal stresses and consequently also for safety.

As Eigenberger has proved in his thesis, in part quite erroneous results are obtained by the lamellae method due to the "divergence" (total force of support $\neq$ total resultant). When using the method presented here, the shape coefficient ζ_{ges} needs to be estimated but it depends only to a minor degree upon the shape of distribution of σ'_N—values between 0 and l.

Disregarding the exceptional case of a very concentrated application of load (see 4.3.2.3), ζ_{ges} may be stated as:

$$\zeta_{ges} = \frac{1}{2}(1 + \zeta_c) = \frac{1}{2}\left(1 + \frac{\hat{\alpha}_z}{2\sin\frac{\alpha_z}{2}}\right)$$

If the factor $\chi_{ges} = \zeta_{ges} \cdot \cos \overline{\alpha}_N$ is also introduced, then the result is $\Sigma|\sigma'_N|$ $= \chi_{ges} \cdot R_{ges}$.

Safety is determined by moment condition, with M as reference point:

$$\eta = \frac{c'_v \cdot r^2 \cdot \hat{\alpha}_z + \Sigma|\sigma'_N| \cdot \mathrm{tg}\, \varphi'_v \cdot r}{R_{ges} \cdot a}$$

or

$$\eta = \frac{c'_v \cdot s \cdot \zeta_c + \chi_{ges} \cdot R_{ges} \cdot \mathrm{tg}\, \varphi'_v}{R_{ges} \cdot \sin \overline{\alpha}_N}$$

where

c'_v	= present cohesion
r	= radius of test circle
α_z	= center angle of test circle
$\Sigma|\sigma'_N|$	= sum of normal stresses
φ'_v	= present angle of friction
s	= length of cord ($s \cdot \zeta_c = r \cdot \hat{\alpha}_z$)
χ_{ges}	= coefficient
ζ_c	= shape coefficient

Therefore it is no longer necessary, as with other methods, to determine $\Sigma|\sigma'_N|$ from the individual lamellae or C'_e, Q'_e, R'_e from several force diagrams, as with the friction circle methods; but safety η may be determined by α_z, $\overline{\alpha}_N$ and R_{ges}.

4.3.2.2 Stratified Soils
At the points of intersection of the individual soil layers and the slip surface, the test mass is divided vertically into slices: $R_{o,1}$, $R_{o,2}$, $R_{u,1}$, $R_{u,2}$ (Fig. 4-15).

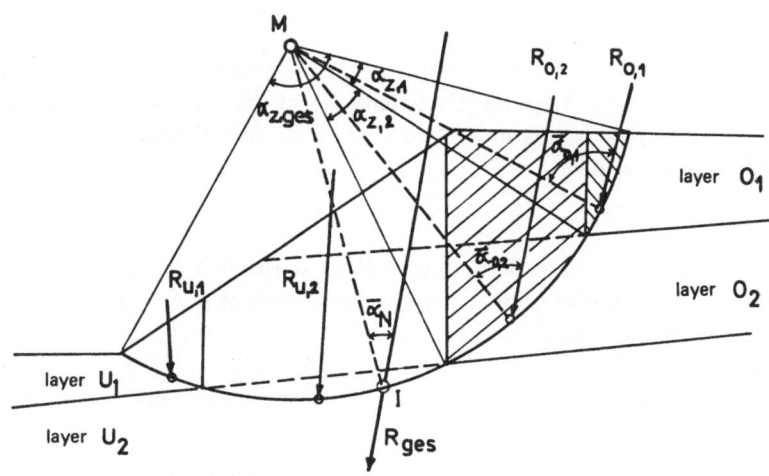

Figure 4-15 Stratified soil, distribution of partial resultants. $R_{o,1}$, $R_{o,2}$, $R_{u,1}$, $R_{u,2}$ are the resultants of external forces, including water pressure acting on slices.

Next, the sum of normal stress $\Sigma|\sigma'_{N,o,i}|$ of the partial resultants $R_{o,i}$, at the right of R_{ges} is determined by the equations shown below. Then the sum of normal stresses $\Sigma|\sigma'_N|$ for the total system is calculated as in 4.3.2.1, and finally the values $\Sigma|\sigma'_{N,u,i}|$ are determined from $\Sigma|\sigma'_N|$.

From the various equations of the known calculation methods the following expression may be found for the test slices o,i at the right of total resultant R_{ges}:

$$R_{o,i} > \Sigma|\sigma'_{N,o,i}| > R_{o,i} \cdot \cos\overline{\alpha}_{o,i}$$

Upper Slices
With the simplified method it is sufficiently accurate if the following expression is chosen:

$$\Sigma|\sigma'_{N,o,i}| = \frac{1}{2}(1 + \cos\overline{\alpha}_{o,i}) \cdot \zeta_{o,i} \cdot R_{o,i}$$

Based upon the minor dependency of the shape coefficient on the shape of the stress distribution, $\zeta_{o,i}$ may again be calculated by:

$$\text{Slice 1:} \quad \zeta_{o,1} = \frac{1}{2}\left(1 + \frac{\hat{\alpha}_{z,1}}{2 \cdot \sin\dfrac{\alpha_{z,1}}{2}}\right)$$

If we define

$$\chi_{o,1} = \frac{1}{2}(1 + \cos\overline{\alpha}_{o,1}) \cdot \zeta_{o,1}$$

this simplifies the equation to

$$\Sigma|\sigma'_{N,o,1}| = \chi_{o,1} \cdot R_{o,1}$$

The same calculation process is valid for slice 2.

Regarding the total system, it may be said that the coefficient χ_{ges} will normally always be larger than the coefficients $\chi_{o,i}$ of the slices at the right of the total resultant, but this only means that the upper part of the test mass is supported by the lower one.

Lower Slices
Therefore, the coefficients of the lower slices at the foot of the slope (left of R_{ges}, subscript u) receive normal stresses on account of their partial resultants $R_{u,i}$ and also on account of the upper layers.

Share from $R_{u,i}$:
For this share the assumption $\chi_{u,i} = \chi_{ges}$ represents a fair approximation. It is then true that:

$$\Sigma|\overline{\sigma'_{N,u,i}}| = \chi_{ges} \cdot R_{u,i}$$

Share from $R_{o,i}$.

For the total system we arrive at:

$$\Sigma|\sigma'_N| = \chi_{ges} \cdot R_{ges}$$

The sum of the normal stresses of all slices obtained by the above equations yields, as a rule, a smaller value than the normal stress of the total system $\Sigma|\sigma'_N| = \chi_{ges} \cdot R_{ges}$.
The difference is

$$\Sigma|\Delta\sigma'_N| = \chi_{ges} \cdot R_{ges} - \Sigma(\chi_{ges} \cdot R_{u,i}) - \Sigma(\chi_{o,i} \cdot R_{o,i})$$

This value $\Sigma|\Delta\sigma'_N|$ is distributed proportionally to the values of the partial resultants $R_{u,i}$:

$$\Sigma|\Delta\sigma'_{N,u,i}| = \Sigma|\Delta\sigma'_N| \cdot \frac{R_{u,i}}{\Sigma|R_{u,i}|}$$

Hence the lower slices are affected by:

$$\Sigma|\sigma'_{N,u,i}| = \Sigma|\overline{\sigma'_{N,u,i}}| + \Sigma|\Delta\sigma'_{N,u,i}|$$

If there is only one partial resultant R_u, situated at the left of R_{ges}, the determination of $\Sigma|\Delta\sigma'_N|$ is not necessary. Then it is easier to determine $\Sigma|\sigma'_{N,u}|$ from:

$$\Sigma|\sigma'_{N,u}| = \chi_{ges} \cdot R_{ges} - \Sigma(\chi_{o,i} \cdot R_{o,i})$$

Safety is again determined by the moment condition in relation to M

$$\eta = \frac{\sum_{1}^{n}[c'_{v,i} \cdot \dfrac{r \cdot \hat{\alpha}_{z,i}}{s_I \cdot \zeta_{c,i}} + \Sigma|\sigma'_{N,i}| \cdot tg\ \varphi'_{v,i}]}{R_{ges} \cdot \sin\ \overline{\alpha}_N}$$

In the exceptional case that one value $\chi_{o,i}$ is larger than χ_{ges}, $\chi_{o,i} = \chi_{ges}$ has to be selected; this will yield results that are on the safe side. Disregarding this exception, the values $\chi_{o,i}$ will almost always be smaller than the "true" values. This grouping has a purpose, because the critical test surface (i.e., slip surface) will avoid the sound layers in the lower zones and cut steeply through them in the upper zone.

4.3.2.3 Concentrated Loads
If forces act only upon part of the slide mass, then the resulting stresses are not distributed over the whole length of the test surface. Therefore ζ, as related to such loads, may not be determined by $\alpha_{z,ges}$. In this case $\Sigma|\sigma'_{N,i}|$ is calculated by dividing the total resultant R_{ges} in two (or more) resultants R_1 and R_2, by estimating the distribution range of stresses. The stresses are determined separately for R_1 and R_2, and then superimposed.

In the case illustrated by Fig. 4-16 it was definitely assumed that R_1 is distributed only from A to C, and R_2 only from B to D:

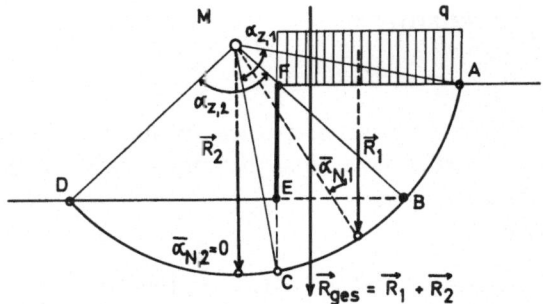

Figure 4-16   Concentrated application of load.

$$
\begin{array}{ll}
\qquad\qquad R_1 & \qquad\qquad R_2 \\
\zeta_1 = f(\alpha_{z,1}) & \zeta_2 = f(\alpha_{z,2}) \\
\chi_1 = \zeta_1 \cdot \cos \overline{\alpha}_{N,1} & \chi_2 = \zeta_2 \cdot \cos \overline{\alpha}_{N,2} = \zeta_2 \cdot 1.0 \\
\Sigma|\sigma'_{N,1}| = \chi_1 \cdot R_1 & \Sigma|\sigma'_{N,2}| = \chi_2 \cdot R_2 \\
\end{array}
$$

$$
\Sigma|\sigma'_N| = \Sigma|\sigma'_{N,1}| + \Sigma|\sigma'_{N,2}|
$$

A division into $R_1 = q +$ weight of soil $ABEF$ and $R_2 =$ weight of soil $BCDE$ only needs to be performed in the case of $R_1 \geq 2/3\ R_2$.

4.3.2.4   Simplification with Pore-Water Pressure (Quick Lowering of Water Table)

Another simplification on the safe side is possible in the presence of pore-water pressure. Immediately after lowering the water table, in Fig. 4-17 from $WSP.1$ to $WSP.2$, the course of the ground-water table will correspond in dense soil to the line $FGHI$. For

$\gamma =$ the unit weight of moist or water-saturated soil

$\kappa_u =$ the pore-water pressure coefficient ($\kappa_u = \gamma_w/\gamma$)

A valid approximation is $W_h = \kappa_u \cdot \gamma \cdot [(h + \Delta h)/2] \cdot h$. From G,A, and W_h results W, and the resultant R is obtained. The calculation should then proceed after 4.3.2.1.

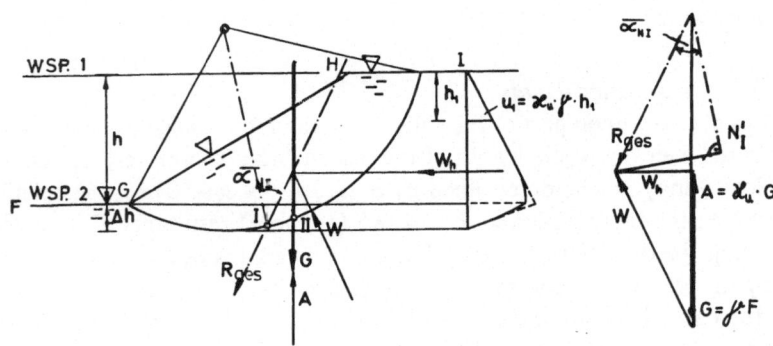

Figure 4-17   Horizontal water pressure during quick lowering.

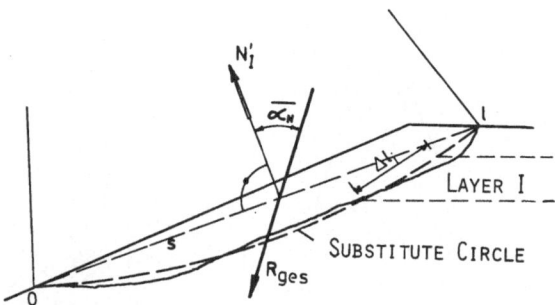

Figure 4-18   An extended slope.

4.3.2.5   *Extended Test Surfaces of Any Shape*

For extended test surfaces of any shape it may be assumed in the approximation that the direction N'_I is vertical to the cord s. The shape coefficient is either stated as $\zeta_{ges} = 1.0$ or estimated with the aid of a "substitute" circle (Fig. 4-18). The test surface may also consist of any combination of circular-cylindrical, logarithmic and other surfaces.

$$\chi_{ges} = \zeta_{ges} \cdot \cos \overline{\alpha}_N$$

$$\Sigma|\sigma'_N| = \chi_{ges} \cdot R_{ges}$$

With stratified soils the division into slices $\Sigma|\sigma'_{N,i}|$ is performed exactly as for circular test surfaces (see 4.3.2.2). Safety is calculated from

$$\eta = \frac{\Sigma[c'_v \cdot \Delta\, l_i + \Sigma|\sigma'_{N,i}| \cdot \mathrm{tg}\ \varphi'_{v,i}]}{\Sigma(R_i \cdot \sin \overline{\alpha}_{N,i})}$$

where

R_i = partial resultant
$\overline{\alpha}_{N,i}$ = angle of the normal to test surface and R_i
Δl_i = length of test surface in layer i

4.3.3   **Calculation of Slope Safety after Eigenberger's Simplified Method — Examples**

4.3.3.1   *Cohesionless, Homogeneous Slopes (Fig. 4-19)*

A simplification of the approximated force diagram after Eigenberger would be as follows.

$R'_{e,I}$ and N'_I apply at I. Safety η is then determined from the moment condition, under the presumption that the absolute value of normal strengths in *II* corresponds to the approximated value N'_I.

The course of the calculation is as follows. A random test circle is assumed, e.g., $r = 10$ m. The resultant of external forces by way of slices or geometric planes (triangle, circle segment, etc.) is determined: $R_{ges} = G = 0.702$ MN

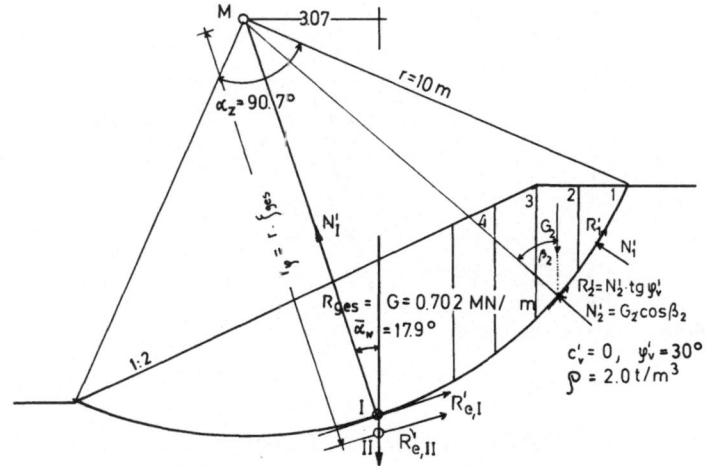

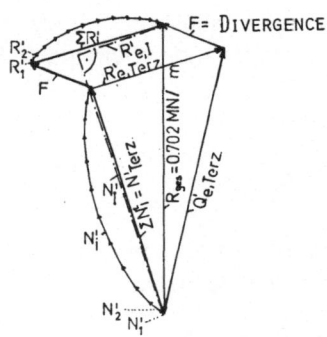

Figure 4-19 Homogeneous slope without cohesion; -.-.-.-.-.- symbolizes approximated force diagram to determine cohesion N'_i, after Eigenberger; ———————— symbolizes the erroneous force diagram (erroneous conclusion!), as assumed in Terzaghi's lamellae method. $Q'_{e,Terz}$ ($\neq R_{ges}$) = force of support taken into account, after Terzaghi.

per square meter. The intersection between R_{ges} and slide circle (I) is found and connected with circle center (N'_I). The angle between R_{ges} and $N'_I(\rightarrow \overline{\alpha}_N)$ is determined: $\sin \overline{\alpha}_N = 0.307$ and $\cos \overline{\alpha}_N = 0.952$. Now the shape coefficient ζ_{ges} is calculated to determine the sum of normal stresses:

$$\zeta_{ges} = \frac{1}{2}(1 + \zeta_c) = \frac{1}{2}\left(1 + \frac{\hat{\alpha}_z}{2\sin\frac{\alpha_z}{2}}\right)$$

where

$\alpha_z = 90.7°$
$\zeta_{ges} = 1.056$
$\chi_{ges} = \zeta_{ges} \cdot \cos \overline{\alpha}_N \cong 1.0$

By the following equation the safety coefficient may be determined ($c_v' = 0$):

$$\eta_1 = \frac{X_{ges} \cdot R_{ges} \cdot \text{tg } \varphi_v'}{R_{ges} \cdot \sin \overline{\alpha}_N} = \frac{1.0 \cdot 0.702 \cdot 0.577}{0.702 \cdot 0.307} = 1.88$$

The improved Eigenberger method yields a safety coefficient $\eta_2 = 1.895$.

4.3.3.2 Cohesive, Homogeneous Slopes (Fig. 4-20)

Here the course of the calculation is the same as in 4.3.3.1. Safety η_1 is calculated as follows:

Cord length: $s = 14.4$ m

$$r_c = r \cdot \zeta_c = r \cdot \frac{\hat{\alpha}_z}{2 \sin \dfrac{\alpha_z}{2}} = 11.13 \text{ m}$$

$$\eta_1 = \frac{c_v' \cdot s \cdot \zeta_c + X_{ges} \cdot R_{ges} \cdot \tan \varphi_v'}{R_{ges} \cdot \sin \overline{\alpha}_N}$$

$$= \frac{0.020 \cdot 14.4 \cdot 1.113 + 1 \cdot 0.702 \cdot 0.25}{0.702 \cdot 0.307} = 2.30.$$

The improved Eigenberger method yields a safety coefficient $\eta_2 = 2.30$.

4.3.3.3 Cohesionless, Stratified Slopes (Fig. 4-21)

At the zone of division between the layers the slide body is vertically divided into single slices (see Fig. 4-15).

Course of calculation:

$$\alpha_{z,ges} = 55° \rightarrow \zeta_{ges} = 1.020$$

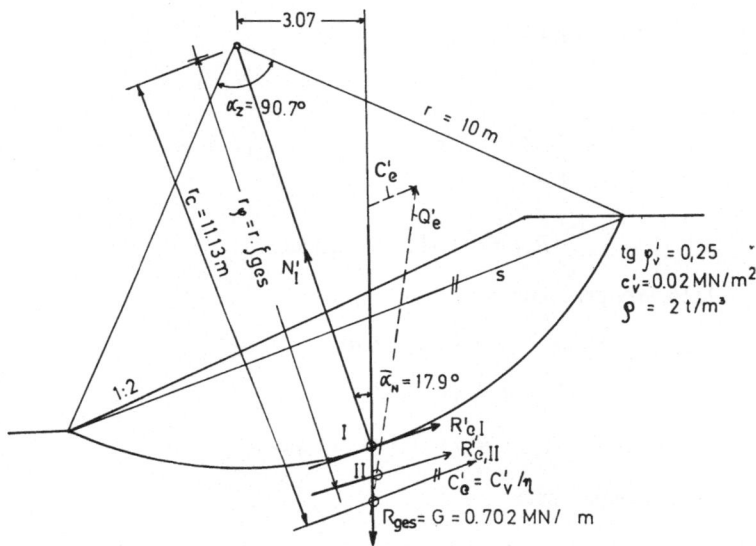

Figure 4-20   Homogeneous slope with cohesion.

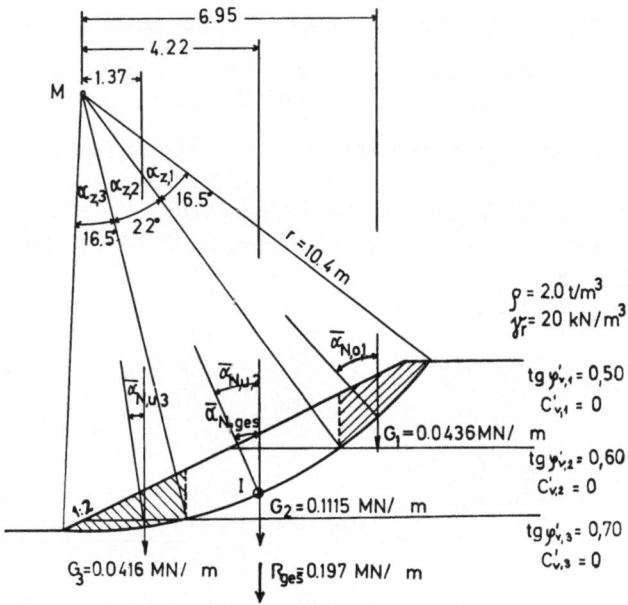

Figure 4-21 Stratified slope without cohesion.

$$\sin \overline{\alpha}_{N.ges} = 0.406; \cos \overline{\alpha}_{N.ges} = 0.914$$

$$\chi_{ges} = \zeta_{ges} \cdot \cos \overline{\alpha}_{N.ges} = 0.932$$

$$\Sigma |\sigma'_N| = \chi_{ges} \cdot R_{ges} = 0.932 \cdot 0.1967 = 0.1834 \text{ MN/m}^2$$

Distribution of partial resultants as in 4.3.2.2.

$G_1 = 0.0434$ MN per meter

$$\sin \overline{\alpha}_{N,o,1} = 0.669; \cos \overline{\alpha}_{N,o,1} = 0.743$$

$$\zeta_{o,1} = \frac{1}{2}\left(1 + \frac{\hat{\alpha}_{z,1}}{2 \sin \frac{\alpha_{z,1}}{2}}\right) \cong 1.0; \quad \chi_{o,1} = \frac{1}{2}(1 + \cos \overline{\alpha}_{N,o,1}) \cdot \zeta_{o,1} = 0.872$$

$$\Sigma |\sigma'_{N,o,1}| = \chi_{o,1} \cdot R_{o,1} = 0.038 \text{ MN per meter}$$

$G_2 = 0.1115$ MN per meter

$$\sin \overline{\alpha}_{N,u,2} = 0.406; \qquad \cos \overline{\alpha}_{N,u,2} = 0.914$$

$$\zeta_{u,2} \cong 1.0 \; \chi_{u,2} = 0.932 = \chi_{ges}$$

$$\Sigma |\overline{\sigma'_{N,u,2}} \qquad = \chi_{u,2} \cdot R_{u,2} = 0.104 \text{ MN per meter}$$

$G_3 = 0.0416$ MN per meter

$$\zeta_{u,3} \cong 1.0 \qquad \chi_{u,3} = 0.932$$

$$\Sigma |\overline{\sigma'_{N,u,3}}| = \chi_{u,3} \cdot R_{u,3} = 0.0388 \text{ MN per meter}$$

$$\Sigma|\Delta\sigma'_N| = \Sigma|\sigma'_N| - \Sigma|\sigma'_{N,o,1}| - \sum_{2}^{3}[\Sigma|\overline{\sigma'_{N,u,i}}|] = 0.1834 - (0.038 + 0.104 + 0.0388)$$

$$= 0.0026 \text{ MN per meter}$$

$$\Sigma|\sigma'_{N,u,i}| = \Sigma|\overline{\sigma'_{N,u,i}}| + \Sigma|\Delta\sigma'_N| \cdot \frac{\dot{R}_{u,i}}{\Sigma R_{u,i}}$$

$$\Sigma|\sigma'_{N,u,2}| = 0.104 + 0.0026 \cdot \frac{0.0115}{0.1531} = 0.1059 \text{ MN per meter}$$

$$\Sigma|\sigma'_{N,u,3}| = 0.0388 + 0.0026 \cdot 0.0416/0.1531 = 0.0395 \text{ MN per meter}$$

Safety:

$$\eta_1 = \frac{\sum_{1}^{3}[\Sigma|\sigma'_{N,i}| \cdot \text{tg } \varphi'_{v,i}]}{R_{\text{ges}} \cdot \sin \overline{\alpha}_{N,\text{ges}}}$$

$$\eta_1 = \frac{0.038 \cdot 0.50 + 0.1059 \cdot 0.6 + 0.0395 \cdot 0.7}{0.1967 \cdot 0.406} = 1.38$$

After the improved method, $\eta_2 = 1.37$ is obtained.

4.3.3.4 Cohesive, Stratified Slopes (Fig. 4-22)
Course of calculation (see also 4.3.3.3):

$$R_{\text{ges}} = 0.7365 \text{ MN per square meter}$$

$$\alpha_{z,\text{ges}} = 105.6° \rightarrow \zeta_{\text{ges}} = 1.078$$

$$\sin \overline{\alpha}_{N,\text{ges}} = 0.312; \cos \overline{\alpha}_{N,\text{ges}} = 0.950 \rightarrow \chi_{\text{ges}} = 1.021$$

$$\Sigma|\sigma'_N| = \chi_{\text{ges}} \cdot R_{\text{ges}} = 0.7522 \text{ MN per meter}$$

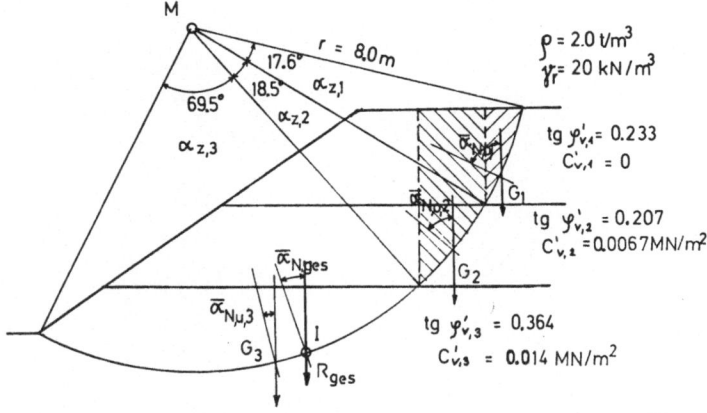

Figure 4-22 Stratified slope with cohesion.

Distribution of partial resultants:

$G_1 = 0.016$ MN per meter

$$\sin \overline{\alpha}_{N,o,1} = 0.899; \cos \overline{\alpha}_{N,o,1} = 0.438; \alpha_{z,1} = 17.6° \rightarrow \zeta_{o,1} \cong 1.0$$

$$\chi_{o,1} = 0.719 \rightarrow \Sigma|\sigma'_{N,1}| = \chi_{o,1} \cdot R_{o,1} = 0.0115 \text{ MN per meter}$$

$G_2 = 0.102$ MN per meter

$$\sin \overline{\alpha}_{N,o,2} = 0.760; \cos \overline{\alpha}_{N,o,2} = 0.649; \alpha_{z,2} = 18.5° \rightarrow \zeta_{o,2} \cong 1.0$$

$$\chi_{o,2} = 0.825 \rightarrow \Sigma|\sigma'_{N,2}| = \chi_{o,2} \cdot R_{o,2} = 0.0841 \text{ MN per meter}$$

$G_3 = 0.6342$ MN per meter

$$\alpha_{z,3} = 69.5° \rightarrow \zeta_{u,3} = 1.03$$

$$\Sigma|\sigma'_{N,3}| = \Sigma|\sigma'_N| - \Sigma|\sigma'_{N,1}| - \Sigma|\sigma'_{N,2}| = 0.6566 \text{ MN per meter}$$

Safety:

$$\eta_1 = \frac{\sum_{1}^{3} [\Sigma|\sigma'_{N,i}| \cdot \text{tg } \varphi'_{v,i} + c'_{v,i} \cdot r \cdot \hat{\alpha}_{z,i}]}{R_{ges} \cdot \sin \overline{\alpha}_{N,ges}}$$

$$\eta_1 = \frac{0.0115 \cdot 0.233 + 0.0841 \cdot 0.207 + 0.6566 \cdot 0.364 + 0.0067 \cdot 2.58 + 0.014 \cdot 9.70}{0.7365 \cdot 0.312} = 1.79$$

The improved method yields a safety coefficient $\eta_2 = 1.78$.

Note that Watari (1973) has published the results of his analyses dealing with the change of the safety coefficient as a function of the aging and the different nature of soils, as well as the significance of soil conditions in relation to slide behavior, and the selection from among the various stabilization measures.

5 Field and Laboratory Investigations

5.1 Field Investigations

In this chapter instructions in textbooks, standard publications, and codes of practice that deserve special attention are repeated and supplemented by a number of aspects that need to be observed.

Recently, it may be frequently noted that exploratory borings are not carried out properly. In my opinion, one of the causes for this unfortunate development lies in the invitation to bid on projects. The clients, in the case of major construction projects, usually state or local authorities, are compelled to give the contract to the lowest bidder. The lowest bidder may not be capable of carrying out such specialized work in the soil of a certain region or may not have skilled workers available who are trained for this type of work. As a consequence, the first period on the construction site is actually a training period and experience is only gained on the job. This may lead to grave errors in the assessment of subsoil conditions due to inexpert exploratory borings. Secondly, as a rule, boring work is contracted at fixed rates per meter. Therefore, boring operations are carried out under time pressure and low quality is the obvious result.

For these reasons, the supervision of the individual stages of boring work by an independent engineer, who is specialized in the specific operations, is a precondition for a technically sound job. To cite an example from my own experience, at a certain construction site only 1.5 to 2.0 m were sunk in the course of one working day while a supervising engineer was present, but 6.0 to 8.0 m were completed during his absence. Inevitably, such a rate of progress can only be achieved at the expense of quality. In Switzerland, as I have learned, regulations make it mandatory that an independent geotechnical engineer always be present on the construction site during boring operations.

5.1.1 Aerial Photography

Efficient and economical measures for the stabilization of landslides cannot be designed without an accurate knowledge of the morphology of the terrain (Blong, 1973, Crozier, 1973).

Aerial photographs, which may be obtained at relatively little expense from cartography institutes, may serve as a first overall orientation. On these photographs, details, such as initial cracks, heavings, etc., are often easier to recognize than by geodetic surveys of the terrain.

It is most important that aerial photographs show the direction to north and the time of the day, the latter in order to be able to interpret the mostly very indicative shadow effects correctly (Colwell, 1980; Dhakharia, 1980; Gupta and Bhandari, 1980; Lueder, 1959).

5.1.2 Geodetic Surveys

Geodetic surveys are, of course, by far more accurate than aerial photographs. From survey results the main direction of a landslide and its type, e.g., whether it is a rotational or translational movement, may be determined. Furthermore, they indicate the position of water outflows, bulges, and ditches on the surface, and from this information conclusions may be drawn regarding water movements in the subsoil and what stabilization measures might be required, such as drainage, the installation of short-circuit conductors, or the construction of supports, such as retaining walls, "Krainer" crib walls, gabionades, anchorings, etc.

Wherever possible, geodetic surveys should be carried out before the stabilization and then at short intervals, in order to monitor character and velocity of slide movements. Geodetic surveys after the completion of the stabilization work serve to confirm the effectiveness of the measures. Stabilization measures may also be carried out so that trespassing of existing boundaries is avoided. Finally, geodetic surveys are the basis for plotting the most indicative longitudinal sections and cross-sections of the slide area.

Geodetic measurements need to be supplemented by the observation of surface movements by means of extensometers installed between benchmarks (Dube *et al.,* 1980; Fukuoka, 1980). These extensometers operate either mechanically or by electric current. In addition, cracks that have developed in buildings may be monitored by measuring devices.

5.1.3 Geological Investigations

An accurate knowledge of the geology of the slide area makes it possible to quickly prognosticate required stabilization measures because, as a rule, similar geological conditions will produce similar types of slides. The above statement is primarily valid within a given region, but knowledge may be gained that is also useful in other regions. In the Vienna area, for example, it is known in which geological era the different clayey-silty deposits originate, and therefore it is possible to draw quite valid conclusions regarding their soil-mechanical properties, even before field and laboratory investigations are undertaken.

Therefore, the cooperation between geoengineering and geology is of decisive significance. The methods described in 5.1.3.1 and 5.1.3.2 may also aid the geologist when carrying out his detailed surveys.

5.1.3.1 Seismic Investigations

The depth of one or several layers of different density may be determined by seismic investigations, as may the depth of a slip surface. The travel of the longitudinal waves is measured at a terminal point by geophones, and from the comparison of the waves traveling on the surface and those reflected by more sound-resistant layers in the subsoil, conclusions may be drawn as to their depth (Bentz, 1961).

This method is relatively simple, but the data should be checked at least from one borehole. In order to assess properly the stratification, attention must be paid to the position of the ground-water table as well as to the fact that, for example, in narrow, V-shaped valleys an echo phenomenon may have a misleading effect. In addition, adjustments are necessary where the surface is strongly undulated or surficial layers have a very inhomogeneous structure, e.g., alternating layers of peat and sand or widely varying thicknesses of fluvial deposits (Bentz, 1961: p. 620).

5.1.3.2 Geoelectric Investigations

These investigations are based on measurements of the electric conductivity of different layers by nonpolarizable sounds. Publications are available that describe the determination of slip surfaces in the subsoil by this method (Fritsch, 1960; Bentz, 1961: p. 718).

This method, however, yields reliable results only if the layers have distincly different structures, expecially chemically and geologically. For example, the zone of division between loose, fractured rock and bedrock is virtually indiscernible because both layers consist of the same material (limestone, granite, etc.), whereby water flow in the soil or rock also plays a major part. In addition, attention must be paid to the presence of natural electric potentials in the soil, for instance, at zones of division or in the vicinity of waste dumps whose margins contain various oxidized, dissolved matters (Bentz, 1961: p. 718).

5.1.4 Soil Exploration by Boring and Extraction of Disturbed and Undisturbed Soil Samples

There is a basic difference between boring work in soft, cohesive soil; stiff, cohesive soil; and cohesionless soil.

A core diameter of at least 15 cm is recommended to recognize inhomogeneities in the soil, such as embedments, slip surfaces, slickensides, or fissures, and to facilitate the installation of the soil sample in the laboratory apparatus. During boring operations the borehole must be cased continuously and the core must be extracted in one piece, disturbing it as little as possible. At each transition from one layer to the next, but at least every 2.50 m, undisturbed soil samples must be taken.

Disturbed bore cores are stored best in 1-m-long wooden boxes; depths of extraction and special data, such as the position of the ground-water table or the swelling of the soil during interruption of boring operations, should be marked on the side of the box, and the location of extraction has to be registered. Color photographs, perhaps with color control strips, are taken of the soil cores and then they are wrapped in plastic foil.

If the soil core is moist, it frequently cannot be recognized immediately whether it consists of clay, clayey silt, sandy silt or silty sand; therefore it is recommended that one cut the surface of the core cylinder along a generatrix, so that an approximately 2-cm-wide plane results. After the core dries out a bit, it is easy to determine the proportion of sand from the degree of disintegration. Immediately after the extraction of the core, strength should be measured by a portable penetrometer and the vane apparatus.

Boring in Soft, Cohesive Soil
Borings are usually sunk by pressure. An approximately 2-m-long core tube is sunk into the original soil. Then the casing tube is pressed in and the core tube is extracted. The casing tube prevents the surrounding soil from falling into the borehole after the core tube has been extracted.

Before extracting the core tube, one must separate the soil core from the natural soil. This is a very sensitive operation requiring great skill and experience on the part of the foreman, and is usually performed by simultaneously rotating and pressing down the core tube. Frequently the casing shoe has a slightly smaller diameter than the upper part of the core tube; another alternative is to place a wire loop in the tube and to cut the core off by tightening. If borings are carried out below the ground-water table, then a valve in the casing cap is required to prevent the water pressure from pushing the soil sample from the core tube while the soil sample is extracted. Creating an air depression in the upper part of the core tube has also been tried, but this requires a more complicated apparatus.

After the core tube has been extracted, the core is pushed out by air pressure and placed in the box. Any unnecessary handling should be avoided so that the core is not damaged. It is not admissible to send part of a core to the laboratory as "undisturbed" soil sample. These have to be extracted at the bottom of the borehole with special extracting tools that are operated like the core tubes but are only 40 cm long and have an inner tin or plastic cylinder that takes up the soil sample. After the casing shoe has been unscrewed, the cylinder is removed and closed at both ends with covers, which are then sealed with paraffin or rubber collars to make sure that the cylinder is absolutely airtight and watertight. Then it is packed in excelsior and shipped to the laboratory. Frequently the core does not completely fill the cylinder and in this case the empty space must also be filled with paraffin.

The soil sample must never be removed from the cylinder before it reaches the laboratory, i.e., it must not be shipped unprotected even though held in place by paraffin. In the case of swelling soils, the covers must be secured by wire to prevent an untimely deformation of the sample.

Easily deformable soil samples, with a water content close to or above the liquid limit, must be enclosed and stored in a manner that prevents consolidation, and they must be transported as quickly as possible from the extraction site to the laboratory. Care must also be taken that during shipment the soil samples are not disturbed by the action of frost.

The removal of the soil sample in the laboratory is relatively simple when plastic cyclinders are used, as these only need to be slit open. If steel cylinders are used, then the sample must be pushed out with a punch; however, this may easily disturb the sample. If the soil has such a high consistency that no cylinder is required, then the soil sample may be extracted without one, but must be covered immediately with paraffin to prevent its drying out.

Before inserting the extracting tool into the borehole, any caving material that entered between core tube and casing must, of course, be removed. Furthermore, the depth of the borehole must be rechecked to determine whether or not swelling due to a possible hydraulic flow has taken place. Such a flow may occur when there is a high hydraulic gradient to the bottom of the borehole. As a remedy, the water level in the borehole must be kept at the same height as the ground-water table. It is also very important to maintain a pressure balance while extracting the core tube. This is achieved by continously pouring that amount of water into the borehole that corresponds to the volume of the core tube plus rods. Special measures are required for cohesionless, finely-grained soils, which will be discussed further on.

Boring in Stiff and Hard Cohesive Soils
What has been said above for soft soils also applies in principle here, but the following details need to be observed. If the core tube cannot be sunk by pressure only, then it must be rotated and a so-called double-core tube has to be used. The inner, actual core tube, into which the core "grows," remains still, while the outer tube rotates independently. With this type of operation it is necessary to flush the space between the two tubes to bring up the boring sludge, but the water must never touch or partially erode the core. It requires a high degree of know-how and sensitivity on the part of the chief operator to make the right choice as to the pressure to be applied, the rate of rotation, and the dose of flushing water. I have seen cases where a bore core of sandy silt was actually burned stiff and transformed into a kind of brick because the driving pressure was too high and flushing insufficient. In another case, inexpert boring compacted silty sand to a degree that the core had a strength of several MN/m^2.

Boring in Cohesionless Soils
Here a core can be extracted with a double-core tube, and this only if the soil contains some silt. If fine, sandy soil is completely cohesionless, then it has such a high angle of internal friction that the core tube cannot be sunk by pressing or by rotation. In this case a star bit with water flushing must be used. The bit compacts the soil slightly and thereby prevents gushing up, while at the same

time flushing makes it possible to sink the bit. To protect the borehole, the casing tube is rammed in as the boring progresses.

With cohesionless soils it is generally not possible to use extracting tools. The investigation of the original soil is carried out by ramming sounds, preferably the standard penetrometer test (SPT)—see also 5.1.5. The best penetrometers are ones that are open at the bottom, because there is a chance that samples of cohesionless, finely-grained soils may also be extracted as they get stuck in the relatively tight (50-mm diameter) tube during ramming.

5.1.5 Measuring Penetration Resistance of Sounds

Sounds are used to supplement a grid of borings (with extraction of soil samples), since soundings are quicker and cheaper to carry out. A great number of sounds are available, which are either rammed or pressed into the soil. With some of them it is possible to measure point resistance and skin friction separately. The bubble-shaped, so-called Koegler or Menard sounds measure penetration resistance as well as horizontal deformation behavior (pressiometer module) of the soil. However, sounding logs must be checked by data obtained from boreholes.

Pressure sounds indicate the pressure required for penetration in MN/m^2. From the change of penetration resistance with depth, conclusions may be drawn regarding soil conditions and the transition from one layer to another. Furthermore, the position of one or several slip surfaces may be determined.

Pressure sounds are mainly used in cohesive and finely-grained cohesionless soils, while the ramming sound is used for the investigation of cohesionless soils.

Internationally, the most widely used sound is the standard penetrometer, which is rammed from the bottom of the borehole. To carry out the so-called standard penetration test (SPT) (Fletscher, 1965), a tube (outer diameter—50 mm, inner diameter—35 mm, length—815 mm), slit along two generatrices, and a driving shoe with a cutting angle of approximately 25° to vertical, is rammed from the bottom of the borehole. The drop hammer sliding on the rod weighs 63.5 kg; the height of drop is 76 cm. From the cleaned bottom of the borehole the tube is first rammed 15 cm into the soil and then the number of drops to achieve a penetration by 30.5 cm is counted. When cohesionless soils contain some silt, it may be possible to extract a soil sample with the tube, but it will be disturbed.

To keep the soil from falling out while the rods are hoisted, a spring locker may be affixed to the driving shoe. In addition, a ball valve at the upper end prevents water in the rods from pushing out the sample while the tool is taken up. When large stones are encountered, it is possible to close the driving shoe with a pointed tip. Ramming with the tip deviates only insignificantly from ramming without it.

When investigations are made in relatively deep boreholes, or under water, it will be expedient to use special sounds (e.g., the Nordmeyer system), with the drop hammer as well as the hoisting and trip gear enclosed in a watertight cylinder, suspended from a rope instead of stiff rods (hammer weight 63.4 kg, height of drops 76.2 cm).

The investigation results are recorded diagramatically, whereby the number of drops required for each 30.5-cm penetration are entered on the abscissa and the corresponding depths, and if possible also the boring profiles, on the ordinate.

5.1.6 Measuring Pore-Water Pressure with a Piezometer

Since landslides mostly occur in cohesive soil layers with a low permeability, only piezometers that register pressure changes with the smallest possible shift of the water volume are suitable for measuring pore-water pressure. Excellent for this purpose are all pressure cells that either transmit the pressure by means of a rather static membrane to a sensitive electric deformation measuring device (Maihak system), or in which the pore-water pressure acting on a membrane is balanced by a measurable counterpressure (Glötzl system). As a rule, pressure cells are installed by boring.

Pore-water pressure gauges should be installed whenever a landslide due to ground failure has to be anticipated, e.g., embankment fills on relatively soft subgrades (Case History 6.2.8 (I)). A sudden steep increase of pore-water pressure in the presumed slip surface may be a signal to stop filling operations. On the other hand, the occurrence of landslides that were evidently delayed by decades may be explained by pore-water pressures that build up slowly (Skempton, 1977b).

If pressure cells are operated from the surface and if the ground-water table is at greater depth, then counterpressure by air pressure has to be applied. This may lead to insignificant inaccuracies.

5.1.7 Deformation Measurements at the Surface and at Different Depths below the Surface

In addition to geodetic surveys, the relative movements of indicative points on the surface may also be monitored by installed extensometers (Fukuoka, 1980; Ter-Stepanian, 1980; Wilson and Mikkelsen, 1978). For this purpose, for example, benchmarks are installed uphill and downhill of the characteristic initial cracks developing at the head of a landslide. The relative movements of the benchmarks can be measured either by permanently connected extensometers, that constantly transmit movements electrically to a control point, or periodically by a person in charge who does it by hand. When a change from a uniform to an accelerating movement is registered then the danger of a landslide is imminent. Figure 5-1 shows how the observation of a slide slope endangering the railway route between Kanaya and Senzu near the Ooi River in Japan was organized, what conclusions were drawn, and what measures taken. Since this railway is one of the main traffic lines of Japan, the period that it

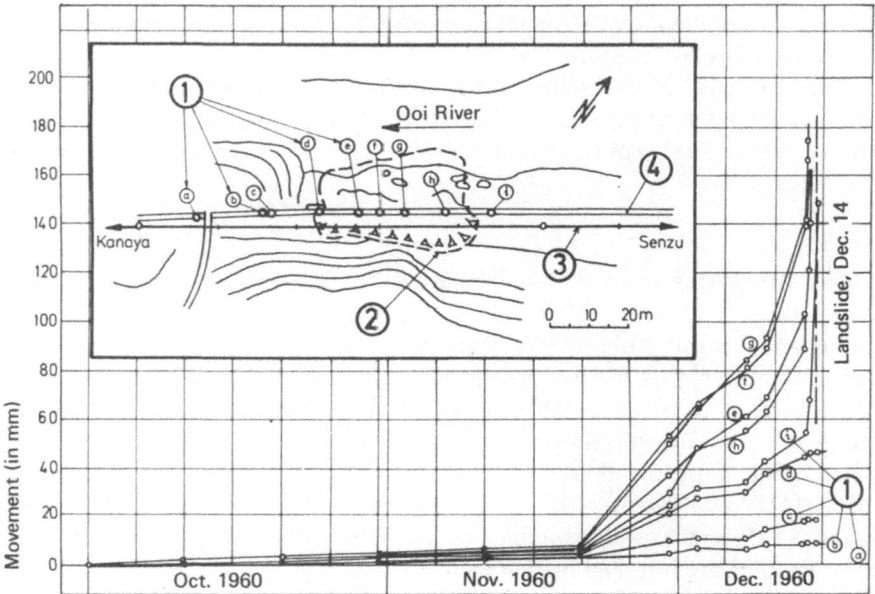

Figure 5-1 Site plan and stress–strain measurements for a landslide above the Ooi River; (1) measuring points *a* through *i*; (2) slide; (3) railway line; (4) retaining wall.

was closed to traffic on account of the anticipated landslide had to be kept to a minimum. Electric stress–strain gauges were installed at nine indicative points, which registered slope movements to a millimeter. At first, movements amounted to 1 to 5 mm per month. From mid-November to early December, movements registered by the measuring points increased to 10 to 100 mm for the 20-day period, and on December 14, 1970, failure occurred. The vital railway traffic was kept up until December 10, and the route was closed for only 10 days, four in anticipation of the landslide and six for clearing away the slide mass, so that economic loss was kept to an absolute minimum.

A simple method to monitor slopes is to set up an uphill and a downhill row of benchmarks, e.g., by driving in long wooden piles that are connected crosswise by electric wires. If a stronger relative movement occurs, the wire snaps and a relay responds immediately. This signal makes it possible to take precautionary measures in time.

In order to monitor the relative movements between the layer above the slip surface and the subsequent layer, tubes are installed in vertical boreholes, as close as possible to the wall of the hole. The profile of these tubes is either round or square, with a prismatic ledge protruding at the inside to protect the inserted inclinometer against revolving around its longitudinal axis. The pendulum in the inclinometer signals any tilting to horizontal by way of electric induction. It is important to monitor these deformations in order to measure any acceleration of the relative movements of the two layers. The extent to which such displacements are possible is demonstrated by Case Histories 6.1.1

(IV) and 6.1.2 (I). For example, when constructing shafts that intersect slip surfaces, the soil deformation to be anticipated may be predicted and appropriate precautions taken.

5.1.8 Measurement Techniques for Electric Soil Potentials, pH-Values, and Redox Properties
by F. Hilbert

5.1.8.1 Measurement of Soil Potentials
In principle, two types of electric soil potentials are measurable: redox potentials (7.2.2.5) and electric potential gradients caused by flow of electric currents in the soil.

(A) Measurement of Redox Potentials. Redox potentials depend upon the ratio of concentration of oxidizing substances to reducing substances in the pore water. A nonreacting electrode (platinum, gold, or—less effective but cheaper—graphite) and a so-called reference electrode (e.g., a mercury-mercurychloride electrode, commonly sold as a calomel reference electrode, or a so-called Thalamide Electrode) are required for these measurements. Figure 5-2 shows the experimental setup. Only high-impedance millivoltmeters, sold as pH-meters, may be used for measuring soil potentials, because all other voltmeters have an input resistance that is too low. The two electrodes should be placed close to each other, but should not be touching. The soil must be well

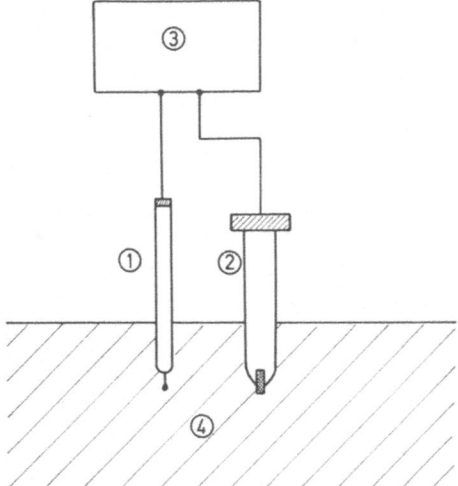

Figure 5-2 Apparatus to determine the redox-potential of the soil; (1) platinum electrode (casing of glass or plastic with an extruding platinum wire); (2) commercial reference electrode; (3) pH-meter (internal resistance $> 10^{12}\ \Omega$); (4) soil, well-soaked with distilled water, into which the two electrodes are pressed.

soaked with distilled water. If the calomel reference electrode saturated with potassium chloride is used, the redox potential at the measuring point is calculated as $U_{\text{Redox}} = 0.241 - U_M\ [V]$; U_M = measured voltage, and $0.241\ V$ = potential of the calomel electrode filled with saturated potassium chloride solution. Portable battery-operated pH-meters are recommended for use in the field.

(B) Measurement of Electric Soil Potentials. This type of potential is generated by electric current flow in the soil, which causes a potential drop of the soil resistance. The currents, in turn, are mainly caused by the different redox potentials in the soil (7.2.2.5), but may also be generated by a hydraulic water flow (reversion of electroosmosis; see 7.2.2.4.2: hydraulic flow in the capillaries generates an electric potential).

It is not possible to take absolute measurements of these electric soil potentials; only potential differences (i.e., electric tension) may be registered. The reference electrodes used for measuring must be identical, i.e., if both are dipped into a potassium chloride solution, the voltage between them, measured with a pH-meter, must be zero. The measurement setup is illustrated in Fig. 5-3; one measuring point is designated as reference point, and one electrode stays there during the whole measuring process. The second electrode is brought into contact with the soil at the spot where measurements are to be

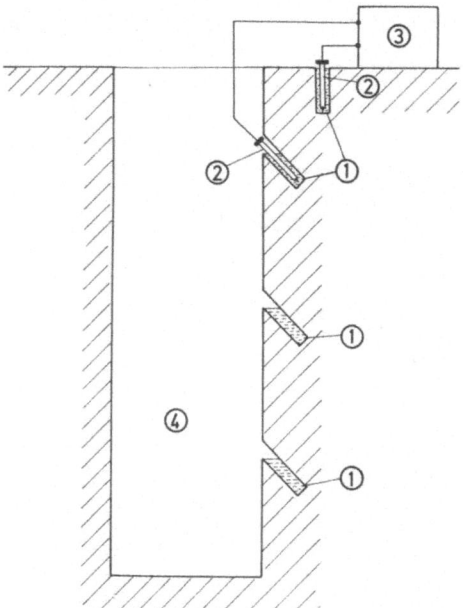

Figure 5-3 Measuring arrangement to determine electric potential differences in the soil; (1) recesses in the side of the shaft i.e. the original soil (after removal of humus), filled with saturated potassium chloride solution; (2) identical reference electrodes; (3) pH-meter registering the potential difference between two reference electrodes.

taken. As in (A), the soil must be well soaked and soft, so that a good electric contact is made between the diaphragm of the reference electrode and the soil material. However, contrary to (A), the soil should be soaked with a saturated potassium chloride solution. If the soil is hard, holes may be drilled and filled with the solution, and the electrode dipped in as shown in Fig. 5-3. For these measurements pH-meters must also be used.

Recently, this method has been considerably simplified. Prinzl (1977) and Pötscher have developed a probe that is inserted into a borehole from the surface. The apparatus is shown in Fig. 5-4. The electric potential difference is measured with calomel electrodes dipped into a potassium chloride solution. The reference point is at the soil surface (electrode 1). Electrode 2, being the measuring electrode, is placed in a pressure tank filled with a potassium chloride solution. Connected to the pressure tank is the probe, which consists of a flexible filling tube, an extendable plastic linkage, and a plastic probe head. This head has holes that connect from the outside to the filling tube at the center; the holes are closed by filtering stones or fritted glass (Fig. 5-5). The probe is brought into contact with the soil at different depths of the borehole, if possible above and below the slip surface, at short distances. The electric contact is made by the potassium chloride solution, which is pressed continuously from the tank (Fig. 5-6) through the filling tube into the probe head, where it slowly seeps through the fritted glass. The applied pressure should be about 0.5 to 1.0 bar higher than the hydrostatic pressure difference between probe head and soil surface. With this apparatus it is relatively easy and inexpensive to measure electric potentials. Of course, the borehole has to be without casing, but in most cases this is possible over a short period.

This type of measurement may also be used for the very important determination of the appropriate distance between short-circuit conductors (6.1.4.4). The method for this application is the following: First one hole is

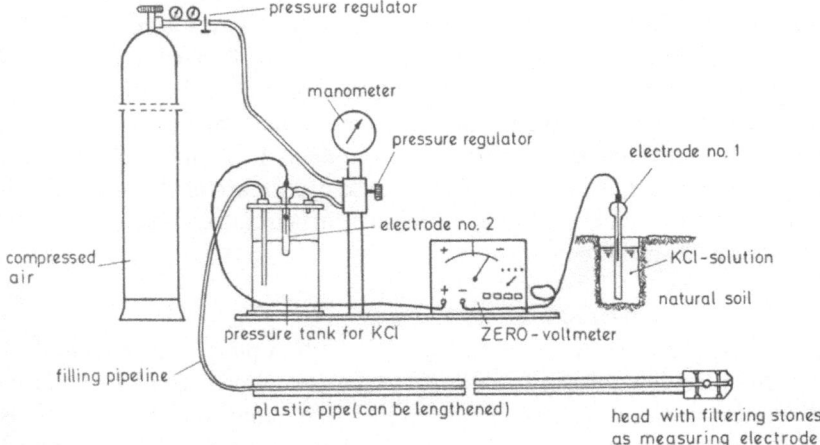

Figure 5-4 Model of measuring apparatus.

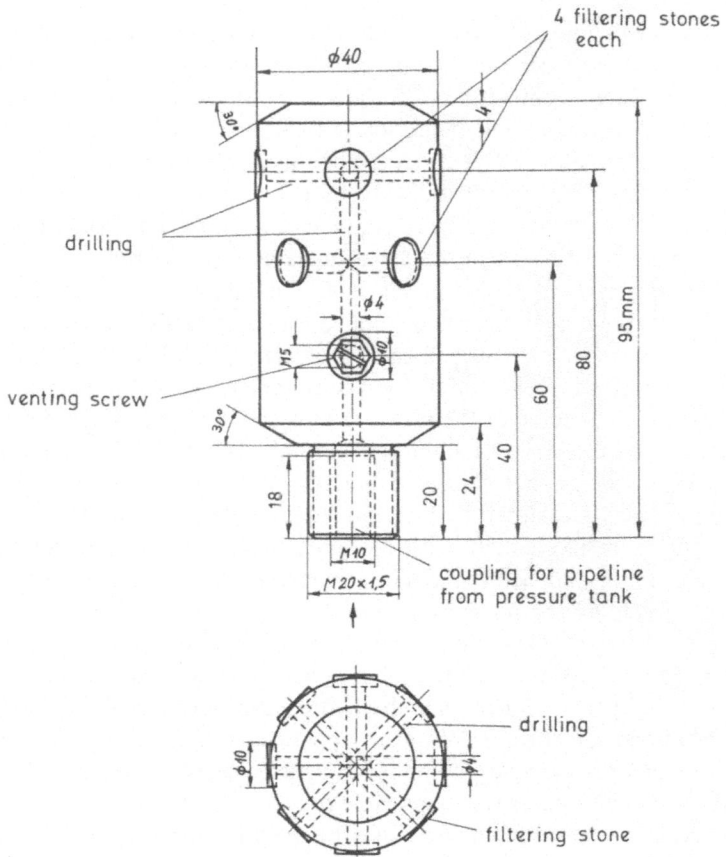

Figure 5-5   Sound head.

bored at the approximate center of the landslide and here the potential difference between the layers above and below the slip surface is measured. Then
the short-circuit electrodes are installed at decreasing horizontal distances and
after several days (3 to 10 days, depending upon soil resistance) the drop of
the potential differences at the borehole is measured while the distance of the
short-circuit conductors is varied, e.g., from 4, to 3, 2, and 1 m. In this way
that distance is determined at which a marked decrease of the original potential difference is achieved. For the grid of short-circuit conductors twice that
distance should be chosen.

5.1.8.2   Measurement of Soil pH-Value

"Combined glass electrodes," designed to be sold as puncturing electrodes, are
well-suited for field measurements. These glass electrodes combine a pH-indi-
cating electrode and a reference electrode. The connection to the pH-meter
(other instruments must not be used) is established via a shielding plug, as
prescribed by the manufacturer in the directions for use.

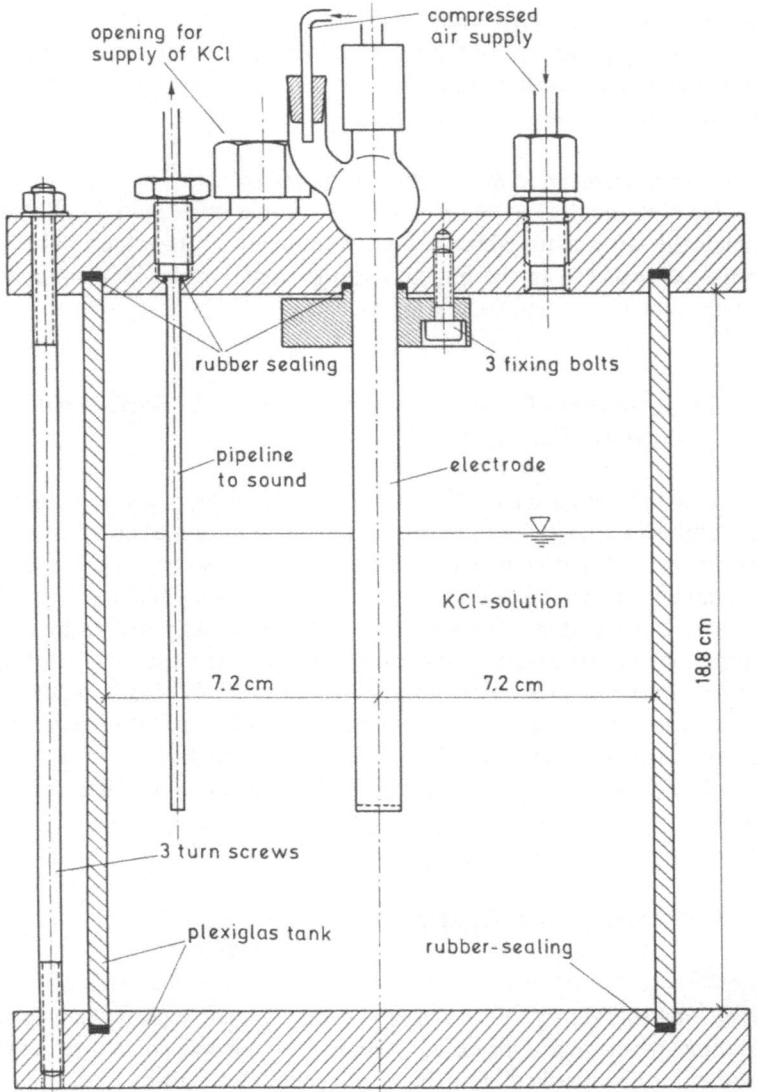

Figure 5-6 Pressure tank.

Measuring the pH-value in the field with a glass electrode is only possible if the soil is so soft that the electrode can be pressed in to a depth of 2 to 3 cm without breaking. The soil must not be softened by soaking as this changes the pH-value.

Field measurements of the pH-value are not exact and at best yield only approximate values of pH-differences between certain parts, for example within one slide area.

More exact values may be obtained by laboratory tests. Undisturbed soil samples are extracted with tridistilled water, filtered and measured in the filtrate. All these operations must be carried out with a very careful exclusion of air. Owing to this requirement, this method is rather complicated and not suited for routine measurements.

5.1.8.3 Measurement of Redox Properties of Soil

The correct method is too complicated for routine investigations. The process as such is the same as for laboratory measurements of pH-values, but the redox potential is measured in the filtrate and the iron(II)/iron(III)-ion concentration determined by chemical analysis.

5.1.9 Measurement of Natural Water Content at the Surface with Radioactive Cobalt

For a quick determination of the natural water content of soil (e.g., before and after stabilizing a larger area or during the construction of embankments), an apparatus is used that contains a radioactive cobalt isotope. The instrument is approximately 40 cm high and of 40 cm in diameter, and has a cylindrical shape with a flat bottom. The intensity of the radiation reflected by the soil is measured, which varies with the natural water content and density of the soil. If the instrument is properly calibrated, these two values may be determined to a good approximation. The use of this relatively expensive instrument is only economical where a large number of measurements have to be carried out. Since it is rather delicate, it should preferably always be operated by the same specially trained engineer.

5.2 Laboratory Investigations

Laboratory operations are described in detail in numerous textbooks, codes of practice, and standard publications, and it may appear superfluous to deal with this subject. Still, there are details which—though known to most experts— are sometimes overlooked, and this negligence may lead to erroneous assessments (Schultze and Muhs, 1967).

5.2.1 Determination of Natural Water Content

The water content w of a soil sample is the ratio between weight of pore water and weight of soil particles, expressed in percentages. The soil sample is dried in a drying chamber at a temperature of $105°C$, until a constant unit weight is achieved. The sample volume needs to be adjusted to the admissible measuring uncertainty, eventual inaccuracies of the scale, as well as the largest grain diameter and water content w.

Numerous methods have been developed to replace the somewhat time-consuming process in the drying chamber. One of these methods is the determination of water content by monitoring the conductivity between two platinum electrodes, embedded in gipsum, which are inserted into the soil sample. Exact experiments carried out at the Graz Technical University have shown that this method for the determination of soil water content is only at all reliable within the very narrow range of about 23–27% water content, and this only if measurements are carried out during an increase in water content. When water content is decreasing, then the values determined are very unreliable on account of the wide range of data obtained (Pötscher, 1977).

Another method, by which the gas pressure generated by a mixture of soil and carbide in a hermetically closed container is measured, has proven unreliable.

5.2.2 Oedometer Test and Determination of Permeability Coefficient

It is well-known that even a carefully performed oedometer test (load test with laterally confined expansion) will usually yield higher values for settlement than actually occur in nature. Furthermore, the time span until a 90% sedimentation is achieved, as calculated on the basis of the oedometer test, will be longer than in nature. The explanation for the greater settlement lies in the fact that a certain relaxation will always take place when the soil sample is removed from the cylinder. During the installation in the oedometer ring, in addition, disturbances are caused at the sides of the sample and at the surface where the cut is made.

Regarding the effect upon the deformation in relation to time, it must be taken into account that the temperature in the laboratory will always be higher than in the soil and the room in which the sample was stored before the test, so that the water enclosed in the pores will often give off air bubbles. These air bubbles narrow the path by which water can seep out during consolidation, reduce k_f (the permeability coefficient) and, in turn, lead to the determination of a longer consolidation period than in nature. Therefore the temperature in the laboratory should be the same as it is in the soil from which the sample was extracted (below a depth of about 1.5 m, soil should be at the mean annual temperature). Another cause for the fact that settling proceeds more quickly in nature are the inhomogeneities that are almost always present, e.g., very small sand lenses that accelerate the dewatering process.

As to the determination of k_f, for different causes mistakes are possible during routine investigations with the available apparati. Some of these mistakes have been discussed already—it is only rarely possible to install the sample without disturbing it, and if no perfect sealing is achieved, especially when dealing with soils with a low permeability, the water will find a path between the sample and the side of the apparatus. The effect of the air bubbles contained in the water as a function of temperature has already been discussed in connection with the oedometer test. In any case, distilled water should be used to avoid a reaction between the alkaline tap water and the glass of the test apparati.

It is generally recognized that results of laboratory tests do not correspond very well to actual soil permeability. It is better to determine this factor by field tests, e.g., by experimentally lowering the water table, though even these tests may not always be successful.

One method, used occasionally where the ground-water table is deep, is to pour water into the soil through a filter pipe and to raise the water table locally; but this may render results inaccurate due to clogging within and around the filter pipe.

5.2.3 Cylinder Compression Test

The uniaxial compression strength of soil samples is the compression strength (q_u) at an unconfined lateral expansion ($\sigma_3 = 0$). It is obtained from cylindrical or prismatic samples with predetermined constant deformation velocity, preferably about 2% of the initial height of the sample per minute. The ratio of height to diameter (length of edge) should be in the range of 2.0:1 to 2.5:1. The test is concluded when either failure occurs after maximum axial force has been surpassed or if, at a large deformation, compression amounts to $\epsilon = 20\%$. The cylinder compression test mainly serves a relatively quick determination of c_u (undrained cohesion), since $c_u = (1/2) q_u$.

5.2.4 Determination of Internal Friction Angle and Cohesion

These values are determined by tests in the triaxial, box shearing, or ring shearing apparatus. In all these tests, loading during the shearing process has to be applied so that the value for peak strength and that for residual strength may be clearly recognized. The method frequently used of stepping up the load to the next value shortly before peak strength is reached (all-around pressure in the triaxial (Bishop and Henkel, 1962) or vertical pressure in the shearing apparati) is not recommended because this method does not yield reliable results either for peak strength or for residual strength. The only test apparatus for the determination of shear strength under different normal stresses is the circle-ring shearing apparatus, which has a theoretical "infinite" shear path. The "Viennese Routine Shear Test" also yields correct shear strength values (H. Borowicka, 1963).

The orientation of the samples in the triaxial and shearing apparatus is of special importance. The sample must be oriented so that the shear plane created during the test lies as far as possible in the slip surface. This means that the sample extracted from the zone of the slip surface is cut and installed so that the slip surface is positioned in the triaxial apparatus at an angle of approximately $45° + \frac{\varphi}{2}$ to horizontal, or lies horizontally in the shearing apparatus. The value φ may either be estimated or predetermined in the box shearing apparatus. This is the only way to determine the shear parameters of the slip surface because these are mostly lower than those of the soil above and below.

5.2.5 Investigations with X-Rays

Methods of investigation with X-rays for the recognition of finely grained crystalline minerals in sedimentary rock have recently come into wider use. In principle, the analysis of minerals by X-rays is based on the fact that every crystalline substance has its characteristic structure of atoms that deflects the X-rays in a characteristic manner. For the purpose of mineral diagnosis the so-called powder method by monochromatic X-rays is used almost exclusively. The powder, that should be as finely grained as possible and in which the minute crystals are oriented at random, is X-rayed. From the lines of interference the spaces of the crystal lattice are calculated by reflection equations, and from these spaces conclusions may be drawn as to the shape and type of the crystal.

The reflected X-rays may be measured by two methods: (1) the photographic method, with radiographic film, and (2) the counter tube or diffractometer method, with Geiger or scintillation counters. When using the first method, the sample revolves around the X-ray; however, today the second method is used almost exclusively and here the Geiger-Müller counter revolves around the radiated soil sample. The interferences are registered automatically.

These radiographical methods are a most valuable assistance, for example, for determining the mineral components of soil layers that take part in the slide processes, whereby the possible presence of readily swelling montmorillonite (bentonite) may be the cause to be looked for. As a rule, the identification of individual clay minerals is rather difficult as many have a similar structure or did not crystallize properly. Therefore, it requires many radiographic tests, and many comparative tests, to analyze the mineral components with sufficient accuracy.

5.2.6 Testing of Models in the Centrifuge

Small-scale soil models that are subjected to centrifugal forces represent the greatest advance made in recent laboratory tests (Lyndon and Schofield, 1978; Smith and Hobbs, 1974). The centrifugal acceleration F (up to 200 g) is the scale coefficient for the dimensions. Processes in function of time are shortened by the square of centrifugal acceleration (F^2).

The effective stresses from the dead weight of the soil (scale coefficient F)

are active during the test. As a consequence of reduced dimensions and time spans, centrifugal tests are superior to large-scale tests, although great experience and a well-tried testing technique are required to carry out stress–strain measurements on the model. The main fields of application of centrifugal tests for analyses are the following:

- Stability of embankments
- Stability of slopes and construction pits in cohesive soil (Beasley and James, 1975; Cheney and Fragaszy, 1980; Lyndon and Schofield, 1970)
- Embankment behavior on soft subgrades
- Settling and load-bearing behavior of sands
- Various interactions between soil and construction, e.g., retaining walls, anchorings, tunnels and pipes in back-filled trenches
- Movements due to earthquakes (Arulanandan *et al.,* 1979)

Centrifugal models, like the models used in the field of hydraulic experiments, are used as an aid for arriving at dimensions and designs that represent technically optimal and economical solutions.

6 Methods for the Stabilization of Landslides

The case histories presented in this chapter were chosen to aid the designing engineer in the selection of the appropriate stabilization method. But as previously pointed out, there exists a wide variety of slide phenomena and ultimately each decision requires a high degree of courage and personal involvement.

Table 6-1 shows the soil coefficients characteristic of the cohesive soils in the Austrian province of Styria, which will be mentioned repeatedly in this chapter.

6.1 The Morphology of the Terrain Remains Unchanged

6.1.1. Moving Layers are Intersected by Construction, but without Arresting the Slide Movement

6.1.1 (I)   Tobacco Factory, Fürstenfeld, Styria, Austria
The long, two-storeyed building without basement, serving as storage building for the tobacco factory in Fürstenfeld, is situated on a terrace above Feistritz Creek. This terrace has a pronounced ledge and from there drops steeply towards the creek. The slope is steepest at a right angle from the longitudinal line of the building, having an inclination of approx 3:4 (vertical to horizontal) or 37°; however, no slide movements had occurred in this area owing to the dense vegetation in the form of trees and bushes (Fig. 6-1).

The already existing scarp turned north and south around the northeastern corner of the building; there was also the shortest distance between building and moving soil. Here the slope had a mean gradient of 28°, being considerably steeper near the head and leveling off towards the foot; this part was only overgrown with grass. Years ago local slides had occurred, some of them starting in the middle of the slope and others in the upper, steep zone; in the process they moved the scarp closer to the building. This caused serious concerns that the scarp would reach the building soon and become a threat to its safety.

Four exploratory shafts reaching to a depth of 5.0 m were sunk to investigate

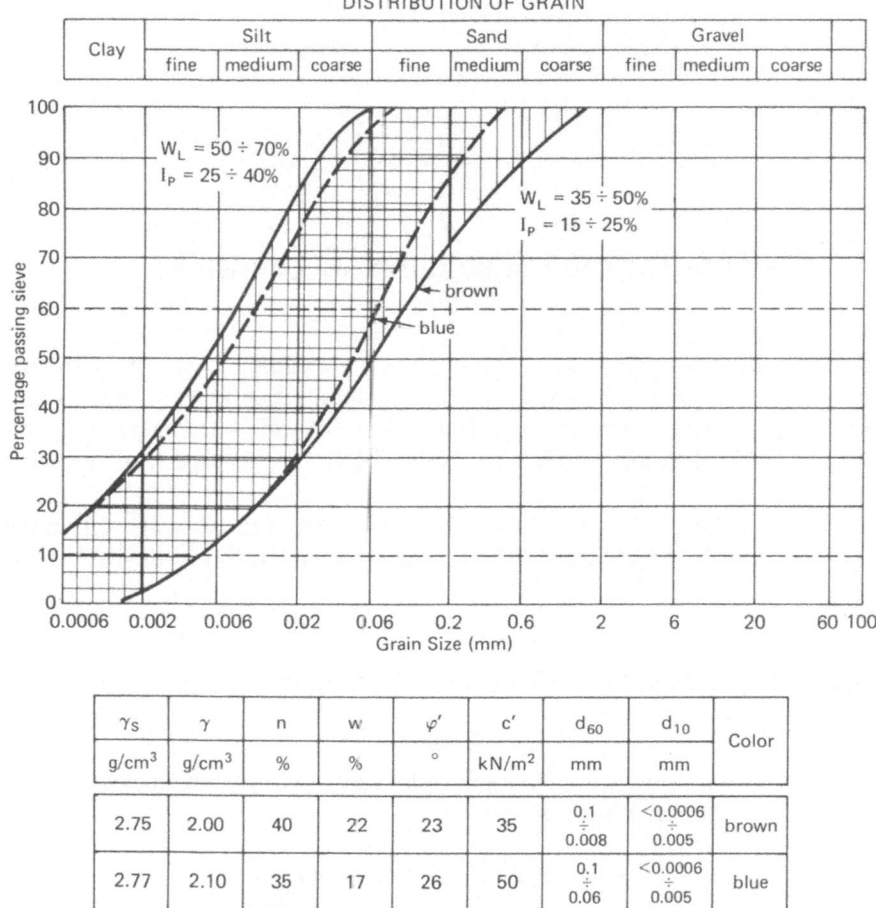

γ_S	γ	n	w	φ'	c'	d_{60}	d_{10}	Color
g/cm³	g/cm³	%	%	°	kN/m²	mm	mm	
2.75	2.00	40	22	23	35	0.1 ÷ 0.008	<0.0006 ÷ 0.005	brown
2.77	2.10	35	17	26	50	0.1 ÷ 0.06	<0.0006 ÷ 0.005	blue

Table of soil-mechanical coefficients of cohesive soils from the tertiary
  (eastern Styria, Austria)
Characteristic d_{10} and d_{60} values of Styrian sandy soils:

$$d_{10} = 0.05\text{-}0.08 \text{ mm}$$
$$d_{60} = 1.4\text{-}1.6 \text{ mm}$$
Geology: Pannon and Sarmat $$d_{60}/d_{10} = 18\text{-}30$$

Table 6-1

stabilization alternatives; they did not yield consistent results regarding the stratification of the soil. Two additional borings, however, did provide satisfactory indications as to subsoil conditions. They showed that the stratification was running at an angle below that of the ground surface; at a depth of about 8.0 m a huge mass of *Tegel* (local term for grey, sandy clay) was encountered. Laboratory tests showed that this *Tegel* was very stiff and of low plasticity; its natural water content was far below the plasticity limit. Since the slope as such did not serve any purpose, it was not necessary to protect it against local slides. In 1970 the building was secured by underpinning, i.e., the construction of a

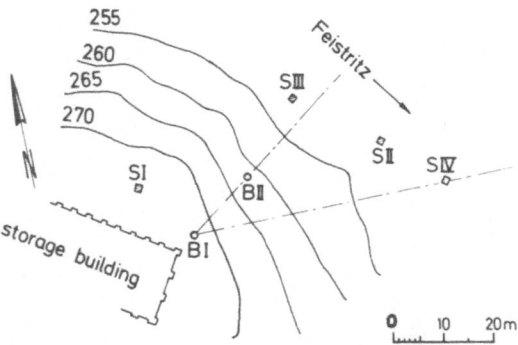

Figure 6-1   Site plan of slide area; S = shafts; B = borings.

footing for the critical, northeastern corner, reaching well into the load-bearing, nonsliding soil layer. Now the slide movements could approach the building without endangering it; they would stop altogether once the slope had been
sufficiently graded down by slides.

The footing (Figs. 6-2 and 6-3) was carried out by constructing two diaphragm-wall panels, joined at a right angle, right next to the walls of the building, which reached to a depth of 8.0 m. This wall was provided with a capping
beam and the flanks were connected with several tension anchors. This construction prevented a relaxation or slump of the soil as the slides moved closer.

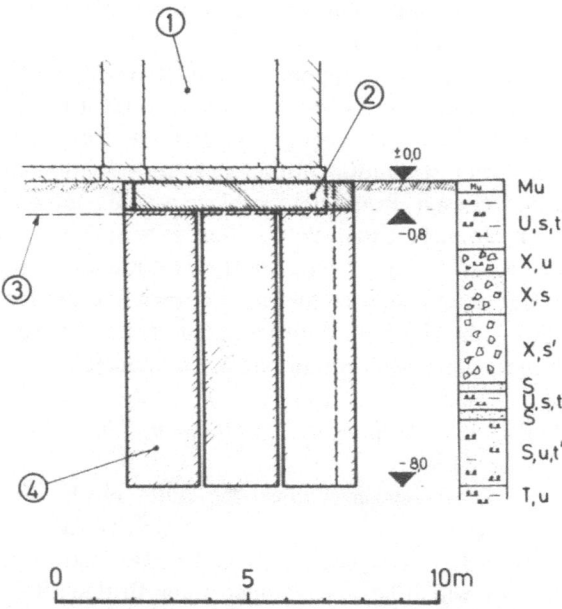

Figure 6-2   Completed stabilization measures, with soil profile; (1) building; (2) capping beam; (3) foundation base; (4) diaphragm wall. (Explanation of symbols in soil
profile—see 6.6.)

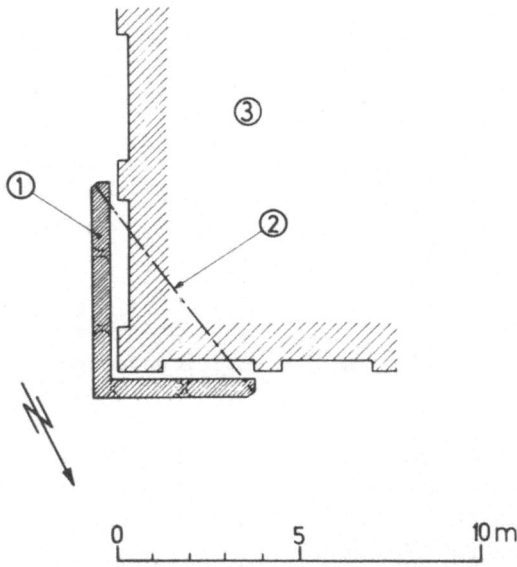

Figure 6-3 Plan of stabilization measures; (1) diaphragm wall; (2) anchoring at level of capping beam; (3) building.

In addition, the slope was stabilized by planting bushes and trees (but none with a trunk diameter of more than 20 cm). Preference was given to plants with deep roots and large water consumption, such as alders, which contribute considerably to natural stabilization of slopes.

The construction of the underpinning was a relatively inexpensive temporary measure, since the building was to be demolished after a few years. The construction of the diaphragm wall could be carried out from the surface of the terrace rather than from the bottom of the creek, representing an added advantage of the method chosen. In order to achieve a stabilization of the slope, a permanent but by far more expensive solution, it would have been necessary first to secure the toe of the slope with riprap or a retaining wall with a proper footing (carried out as an anchored pile wall or anchored diaphragm wall) and then to grade down the slope. In addition, a thorough drainage of the slope, perhaps by horizontal drainage, would have been indicated.

6.1.1 (II) Foundation of the Lueg Bridge, Brenner, Tirol, Austria (Wenzel and Fenz, 1970)

After crossing the Obernberger Valley, the route of the Brenner freeway ascends towards Lake Brenner; it crosses the Sill River at a height of 50 m above the riverbed. This whole section of the freeway had to carried out as a slope bridge on piers, which has been named Lueg Bridge after a nearby little church.

The piers, most of them between 15 and 30 m high, were designed in the shape of elongated hexagons (length—7.0 m, width below roadway axis—1.5 m, at the flanks—0.9 m; Fig. 6-4).

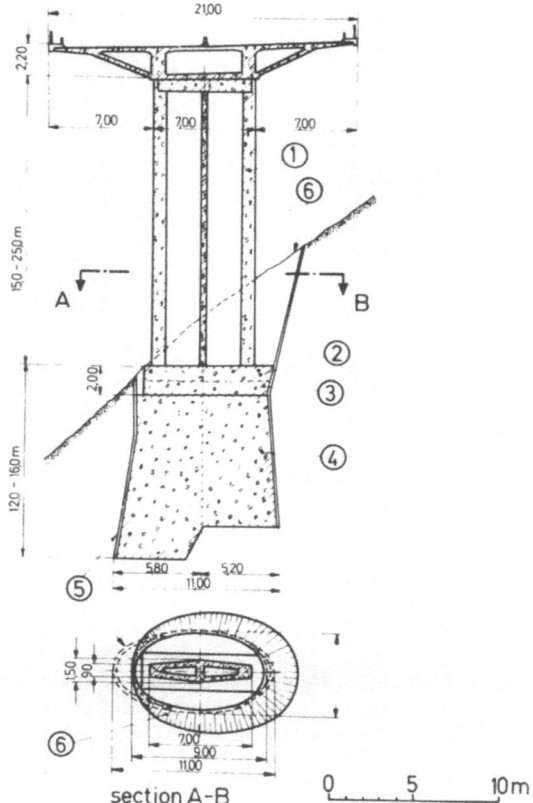

Figure 6-4 Section and plan of pier; (1) B 400; (2) B 300; (3) B 225; (4) B 160; (5) shot-concrete skin; (6) preliminary cut.

The very high piers in the area of the riverbed required two additional profile designs, but these corresponded to the basic design to make it easier to adapt the concrete forms. Slip-forms were used for the construction of all piers. The hexagonal shape met the static requirements of the Lueg Bridge and also looked good.

For one pier the foundation had to be constructed in mica-slate talus in an area where a susceptibility to creep movements could not be definitely excluded (Fig. 6-5). Expert opinion indicated that a creep movement of 1 cm per year had to be anticipated and consequently taken into account in the design. This creep movement would probably reach to the bedrock surface, with an assumed dip of approximately 25°, that is, only about two-thirds of the surface gradient. Therefore the pier foundation had to lie below this zone of division. Naturally, exact data regarding the position of this zone of division were limited and this made it advisable to anticipate possible small movements of the pier foundation in the bridge design. Consequently, provisions were made for an eventual horizontal as well as vertical adjustment of the superstructure by 30 cm. Furthermore, the forces of an eventual slope movement had to be either taken up or

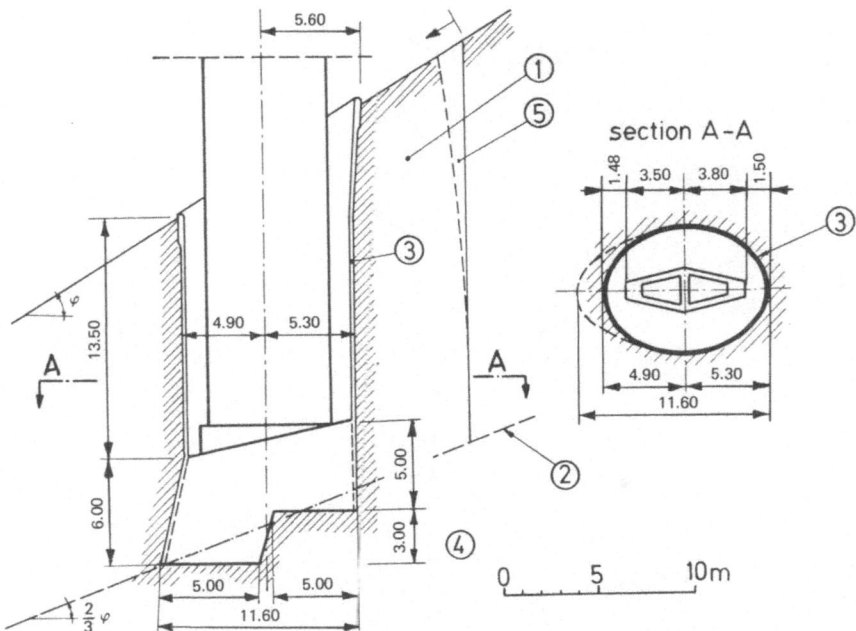

Figure 6-5 Section and plan of pier foundation in critical zone; (1) slope susceptible to creep movements; (2) zone of division; (3) shot-concrete skin; (4) bedrock: quartz-calcium mica slate; (5) course of creep deformation.

transmitted before they reached the pier. In the end, the following solution was found:

(1) The pier foundation was positioned below the zone of division and was given an elliptical shape; it reached a maximum depth of 20 m. A maximum edge pressure of 1 MN/m^2 at the base was foreseen.

(2) The pier above the foundation was protected by an 11.0-m-long and 7.60-m-wide elliptical tube. To make the pier itself able to take up the forces resulting from eventual slope movements was not economically feasible. On the uphill side there was a clearance of 1.20 to 1.60 m between the 7.30-m-long pier and the wall of this hollow, elliptical tube, allowing it to follow a slope movement without touching the pier. As the tube moves with the slope, it is under no additional strain, except in the case of a nonlinear creep profile and a resulting longitudinal bending moment which, however, would have virtually no impact upon the stability of the construction.

The hollow ellipse received first a 20-to-25-cm thick skin of shot concrete; no bracing was required (Fig. 6-6). Using shot concrete had the advantage that no large boring equipment was needed, as it would have been for bore-pile and diaphragm walls, which would have been difficult to maneuver on the very steep slope.

The shot concrete skin was reinforced with two to three layers of steel net and, being relatively thin, was so deformable that the play of forces caused by

Figure 6-6 Pier foundation in critical zone of slope.

the excavation of the construction pit soon came to rest again. Since the concept was based on an arching effect, it was important that the ellipse did not deviate too much from a circle, i.e., that it was not too elongated (Fig. 6-5).

(3) The hectagonal piers were made hollow; at the front facing the slope, 30 cm were added to the width, so that in the case of a horizontal displacement of the pier the superstructure could be brought into its proper position again.

(4) Since any construction pit excavation on a very steep slope entails the danger of disturbing the soil and since, in addition, deep foundations combined with such hollow ellipses are naturally quite expensive, the bridge spans between piers were designed twice as long as for other bridge constructions in the vicinity.

This solution proved to be very effective here and in other similar cases. The construction of a protective shell for the pier, to prevent the pressures due to slope movements from reaching it, made it superfluous to secure the pier itself

by expensive measures, such as vertical and horizontal anchors. In more difficult cases, i.e., where slope movements are more severe, the protective shell
needs to be anchored at the side facing the slope.

6.1.1 (III)   Foundations for Cable-Car Supports and Powerline Poles

The basic concept is the same as for the two bridge foundations described
above and below (6.1.1 (II) and (IV). In this case a sliding, surficial layer of
brown, silty, plastic clay several meters thick, and a subjacent layer of grey,
silty, stiff clay (Fig. 6-7) were encountered. The foundations reaching into the
stiff clay were designed so that the upper, sliding layer could move around
them like water around a bridge pier; this is similar to Case History 6.1.1 (IV);
(Figs. 6-8 through 11).

In many cases it is advantageous to use diaphragm-wall panels; they find a
firm footing in the load-bearing subsoil (ICOS 3a, 1968). This construction
method has the following advantages:

(1)   When carrying out the load-bearing elements with diaphragm-wall
      panels the soil is not disturbed, i.e., neither relaxed nor compacted.
      This means that construction activities will in no case accelerate an on-
      going slide or set off a new slide.
(2)   The design of the foundation can be varied, e.g., a simple, long dia-
      phragm wall, or I-shaped or cross-shaped elements can be used. This
      selection of shapes makes it possible to achieve an optimal adaptation
      of the foundation to the terrain. To be capable of transmitting hori-
      zontal and vertical forces and the existing moments, all panels must be
      provided with steel reinforcements.
(3)   The construction equipment for diaphragm walls is light, easy to
      maneuver, and therefore well suited for work on mostly soft ground.

This method of constructing safe foundations in sliding or slide-prone soils
has been used successfully in many areas and, in my opinion, any other methods would have been more expensive. One possible alternative, for instance, is
vertically and/or horizontally anchored concrete blocks. But these have to be

	< 4 mm	< 0.06 mm	< 0.005 mm	natural water content	plasticity limit	liquidity limit	plasticity index	content		
								SO_3	Fe_2O_3	FeO
upper layer brown silty clay	% 100	% 86.7	% 0.6	% 55	% 23.6	% 63.5	% 39.9	% 0.58	% 3.68	% 0.7
lower layer stiff gray-blue clay	100	89.4	48.3	32	21.4	73.8	52.4	1.56	1.26	3.17

Figure 6-7   Soil-mechanical coefficients above and below slip surface (characteristic
for some areas in central Sicily).

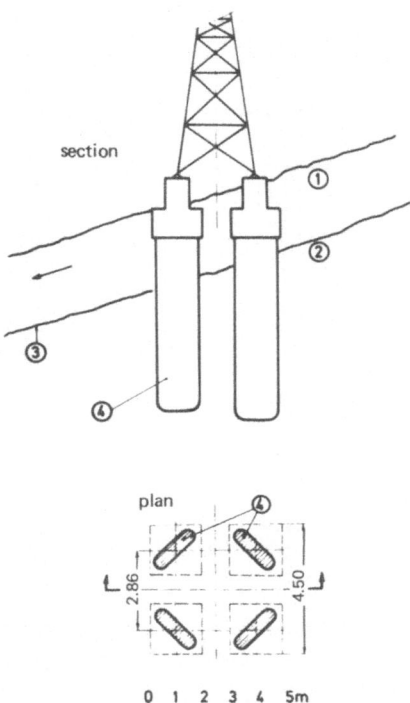

Figure 6-8 Construction with *I*-shaped diaphragm-wall panels; (1) brown, silty clay; (2) stiff, grey-blue clay; (3) slip surface; (4) diaphragm-wall panels.

constructed either in open, braced excavations, or behind protecting sheet-pile or bore-pile walls. The anchors must reach all the way through the surficial slide layer.

6.1.1 (IV)   Foundation of the Limberg Bridge, Franz-Josefs-Railway, Lower Austria (Raschka, 1912)

The Limberg Bridge is the first crossing of a landslide by a railway route, whereby the bridge, constructed from 1910 through 1912, is founded on deep piers.

The section Ziersdorf–Eggenburg of the Franz-Josefs-Railway lies in Lower Austria. The terrain there is hilly, frequently interdivided, and intersected, so that the construction of the railway route necessitated, among others, up to 28-m-high embankment fills. In 1870 the railway route was completed, at that time with only one track, and in 1903 came the construction of the second track. During the one-track operation, the Limberg embankment already showed signs of instability; during the construction of the second track a culvert had to be replaced.

The year 1910 brought heavy precipitation and in autumn it could be observed that the embankment for the downhill track was separating from the uphill half (Fig. 6-12). At first the movement was hardly noticeable and could

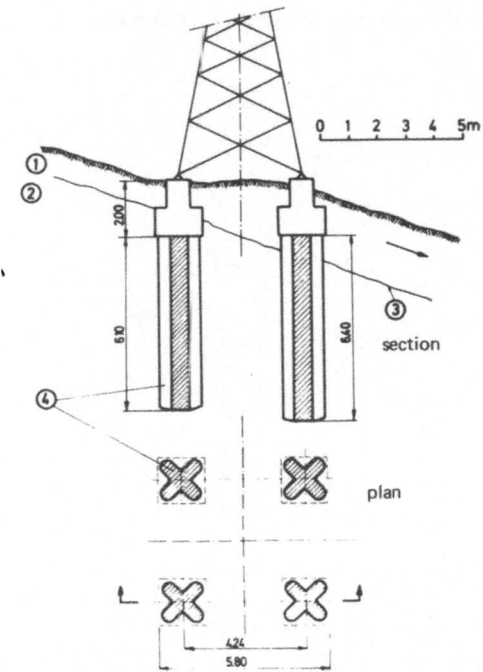

Figure 6-9 Construction with cross-shaped diaphragm-wall panels; (1) brown, silty clay; (2) stiff, grey-blue clay; (3) slip surface; (4) diaphragm-wall panels.

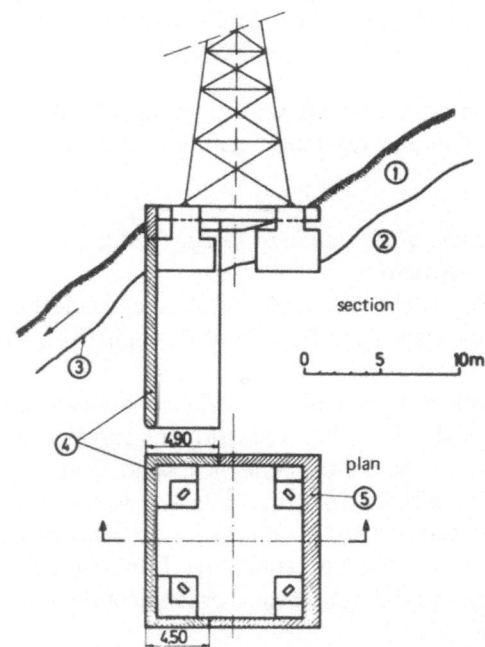

Figure 6-10 Construction with box-shaped diaphragm-wall panels; (1) brown, silty clay; (2) stiff, grey-blue clay; (3) slip surface; (4) stratified sandstone; (5) diaphragm wall; (6) connecting beams.

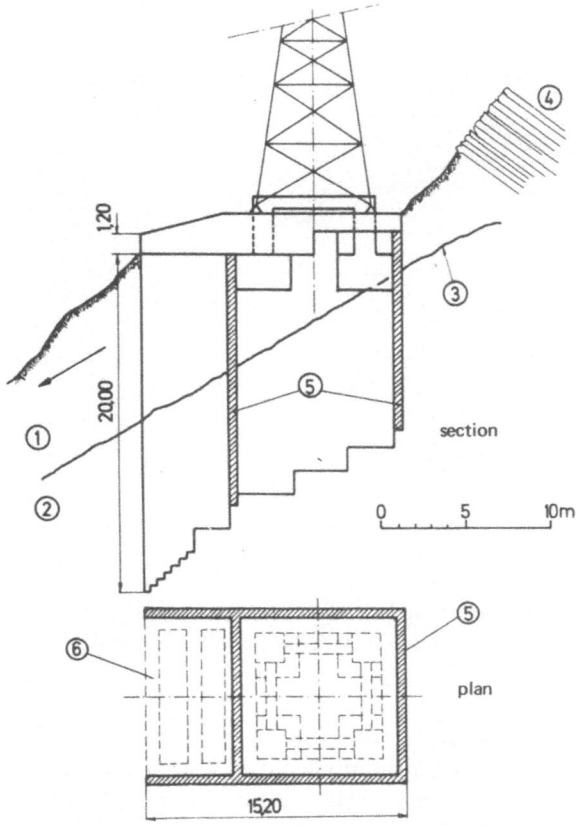

Figure 6-11 Construction with *U*-shaped diaphragm-wall panels; (1) brown, silty clay; (2) stiff, grey-blue clay; (3) slip surface; (4) diaphragm wall; (5) connecting beams.

only be determined with certainty from the widening of the culvert joints (Fig. 6-13). Later, movements of 3 to 4 cm per day were registered. At the same time the soil under the downhill vinyards and fields was heaving up in a bulge, indicating that the whole slope was in motion.

An exploratory shaft that was sunk in winter reached bedrock at a depth of 19 m. Together with the shaft, a number of boreholes were driven down to bedrock, yielding a clear picture of the nature of the soil and its stratification as well as a few indications as to the cause and extent of the landslide. Down to a depth between 5 and 15 m the soil consisted of silt and sandy silt, then followed a layer of stiff clay, a thin layer of gravel, and a sand layer; finally, granite bedrock was reached at varying depths (Fig. 6-14). This bedrock had a steep dip to the valley; its strike was parallel to the line Vienna–Gmünd; it cropped out near one end of the embankment, dropped steeply to 12 m and then evenly to 35 m below ground surface until it reached the other end of the embankment. The deepest borehole reached bedrock only at a depth of 38 m.

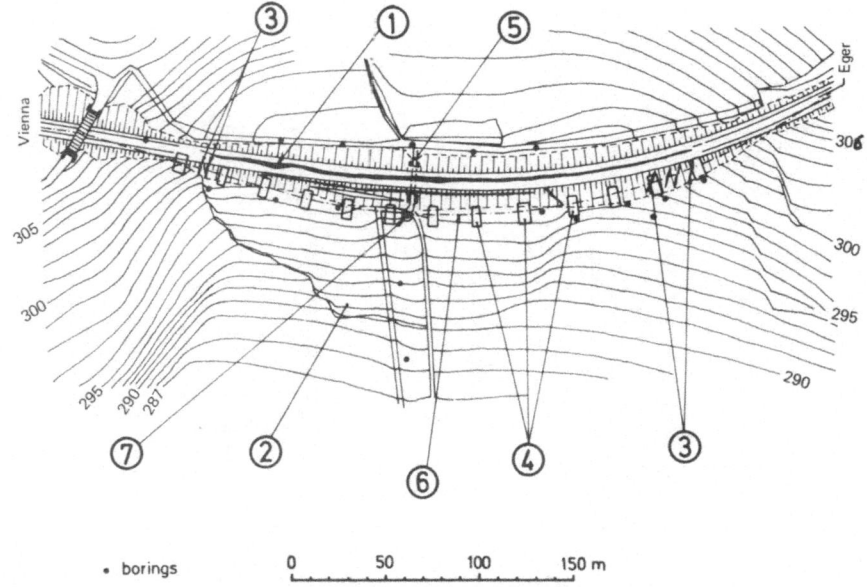

Figure 6-12 Site plan of slide area; (1) initial scarps; (2) bulge; (3) lateral cracks; (4) bridge pier; (5) culvert; (6) new railway route; (7) shaft.

The landslide covered an area of 150 × 50 m. From the observation shaft it could be determined that a slightly inclined slip surface was present at a depth of 6 m, that the soil above this slip surface was greatly disturbed, and that it was at rest below. Gradually, the upper part of the shaft was shifted by 3 m and completely separated from its lower part (Fig. 6-15). Consequently, at this spot the slide layer reached to a depth of 6 m. Assuming that this was the average depth, the sliding mass had a volume of about 80 m^3, at that time the largest known soil movement.

There were the typical characteristics of a landslide: At the crown there were wide, continuous, *transverse* cracks; above the flanks similar *longitudinal*

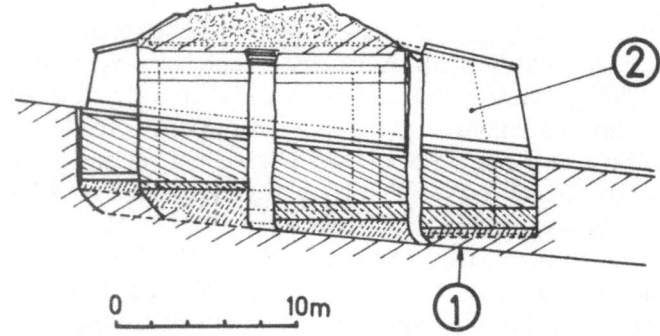

Figure 6-13 Section of arched culvert; (1) slip surface; (2) destroyed culvert.

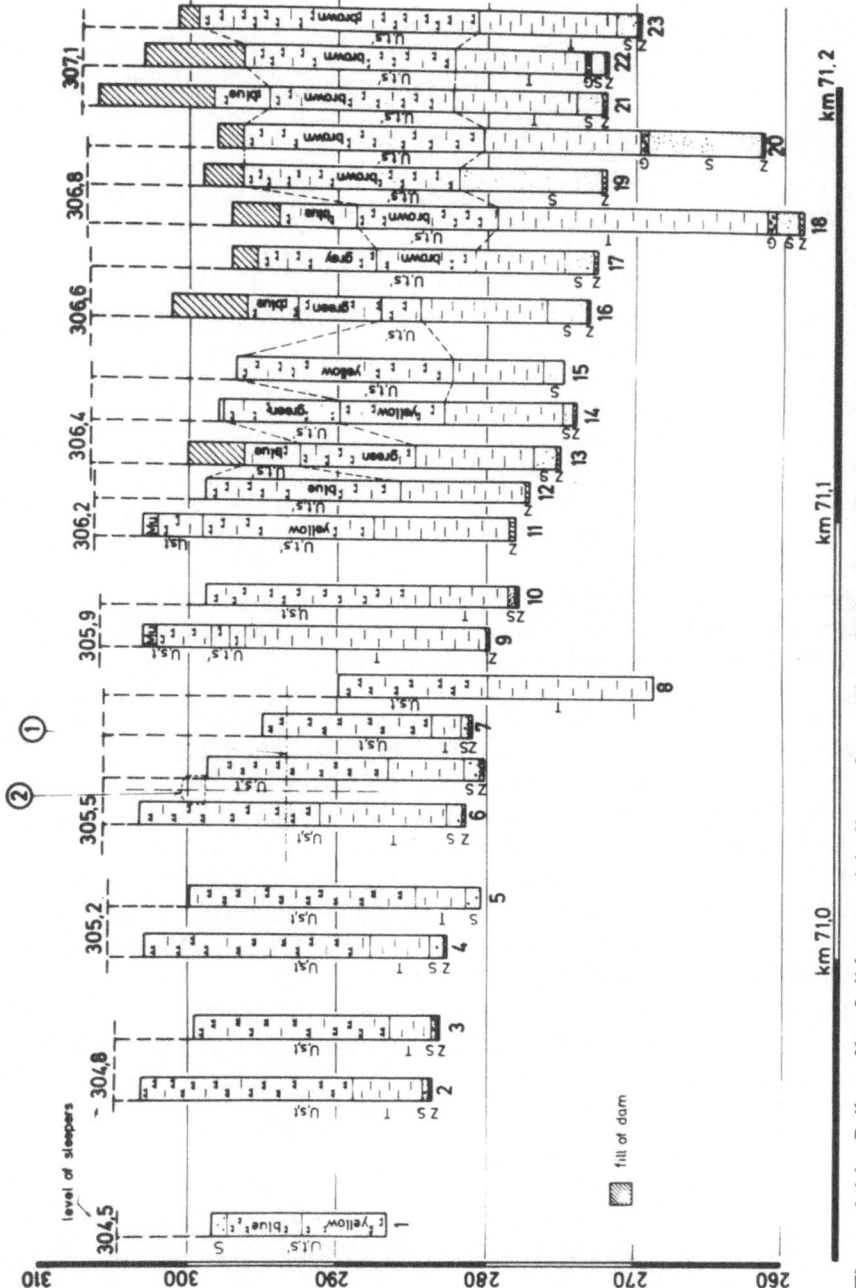

Figure 6-14 Soil profile of slide area; (1) slip surface; (2) arched culvert. (Explanation of symbols in soil profile—see 6.6.)

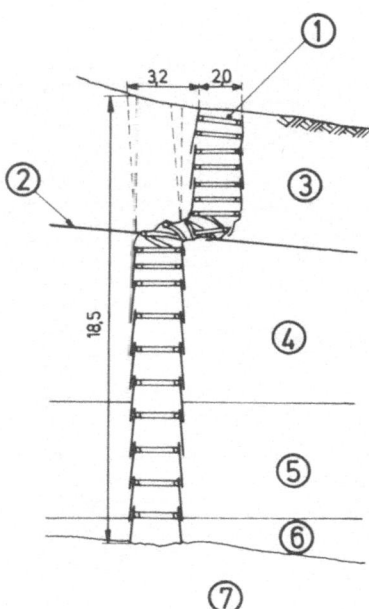

Figure 6-15 Damaged test shaft; (1) test shaft; (2) slip surface; (3) yellow, sandy-clayey silt; (4) yellow, sandy-clayey silt, being stiffer at greater depths; (5) stiff, grey clay; (6) densely deposited yellow sand; (7) bedrock.

cracks and fissures; at the foot an advancing bulge. The subgrade of the slip surface was almost as smooth as a mirror (Fig. 6-16).

From November until May the slide movement amounted to an average of 2.5 cm per day or 0.75 m per month, being fairly uniform. Only after long rainfalls, usually three to four days later, did the velocity of the slide movement increase. Apparently the weight of the embankment, increased by the load and vibration of railway traffic, triggered the slide; but the extraordinary soaking of the soil during the rainy year 1910 also contributed.

The construction of the second track increased the weight of the embankment considerably. The slope carrying the embankment may have been close

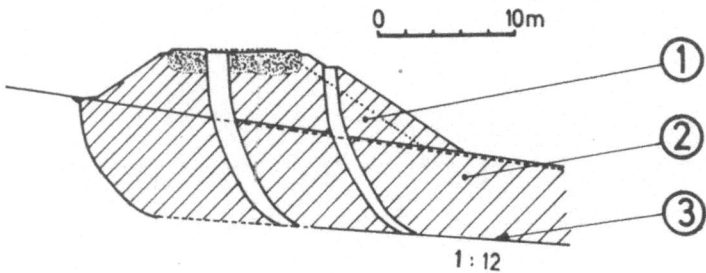

Figure 6-16 Typical section of slide; (1) dislocated embankment; (2) subsoil below embankment; (3) slip surface.

to limit equilibrium already in earlier rainy years, when the strength of the silty soil was reduced by soaking, although its mean inclination was only 1:6 (vertical to horizontal). After 1903, when the load of the embankment was increased, 1910 was the first year with excessive rainfall. Under the increased load the soaked sandy silt could no longer maintain its equilibrium and started to slide; failure had to occur at a depth were the soil had remained drier and firmer, and consequently the slip surface ran right through the middle of the huge mass of silt.

Four remedial proposals were reviewed.

(1) The construction of stone ribs or a retaining wall at the toe of the embankment, strong enough to prevent a further movement of embankment and slope, and permitting a relocation of the displaced part of the embankment to its original position.

(2) Extensive drainage of the whole slope by trenches and ducts.

(3) Relocation of the whole railway route to a safe, nonsliding area.

(4) Replacement of the embankment by a bridge on piers with sufficiently deep foundations (Fig. 6-17).

Of these four alternatives, the construction of a retaining wall proved too expensive on account of the required great foundation depth. As to draining the slope, it could not be assumed beyond a doubt that it would be wholly effective, since the movement of slope and embankment did not even stop during the extremely dry year 1911; even an efficient drainage system has little effect in rich, barely permeable silt. A complete relocation of the railway route was out of the question on account of the prohibitive costs.

Therefore, the only safe and economical solution was to cross the slope on piers, that is, to build a bridge. As railway traffic was not to be interrupted, the bridge could not be constructed on the old route; therefore, all piers and abutments had to be founded next to the embankment and the downhill side was chosen. The piers were set at distances of 20 m and the structure members consisted of girders.

Construction work commenced in April 1911; to prevent further movements, the piers closest to the culvert were founded first. Actually, that time of the year is not the best for construction work, but the successful completion proved

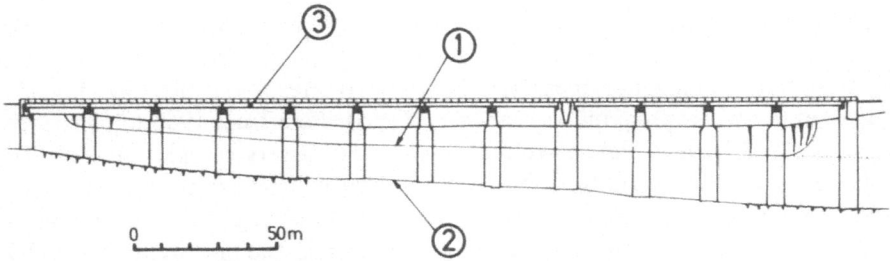

Figure 6-17 Longitudinal section of bridge replacing embankment; (1) slip surface; (2) bedrock surface; (3) bridge.

that it was right to go ahead. The broken culvert came to rest and with it the most critical zone of the landslide. Even though the bracings of the foundation pits for the middle piers were strutted with heavy timber, they were again and again crushed and the upper 6 m of the construction pits pushed downhill by more than 2.5 m. Nevertheless, it was possible to keep the construction pits open, but it became necessary to carry out additional excavations on the uphill side. At the other piers the pressure of the sliding masses was not so strong. The dry weather of 1911 also greatly favored construction work, which was completed in autumn of 1912.

At that time this was the first excavation of pits for the construction of pier foundations with such dimensions; they intersected a sliding mass of 5 to 6 m thickness and continued to depths of up to 25 m below ground surface.

The chosen solution to intersect the sliding slope with piers reaching into a load-bearing layer, whereby the sliding soil could bypass the piers like a river flowing around bridge piers, strikes one as being quite modern (see Case Histories 6.1.1 (II) or (III)). Nowadays the pier foundations would have been constructed from the surface, using bore-pile or diaphragm walls, thus eliminating the need for deep, open excavations.

6.1.2   Stabilizing Moving Soil Layers

6.1.2 (I)   A Landslide in Turkey (Hamdi, 1969)
This case history deals with an old landslide. After a relatively long period of rest, in 1963 a series of landslides started that dislocated a railway line in Turkey and destroyed an asphalt road (Fig. 6-18).

After each slide the railway tracks were relocated to their original position in order to keep up traffic. At the same time a row of measuring points was installed along the route and these were constantly observed. During a period of 4 years a total displacement by 4.68 m was registered there; another row of measuring points, set up at a distance of 25 m from the railway tracks, indicated a displacement by 22.75 m. The movement of the riverbank during the prior 10 years amounted to about 20 m. Besides the railway line and the road, the slide movements also started to endanger residential buildings situated close to the flanks of the slide.

A series of borings and laboratory investigations clearly indicated that additional explorations and long-term observations were required to determine the nature of the movements before a decision could be reached regarding the appropriate stabilization measures. Based on the results of the 1963 investigations, a precise program was set up and carried out from 1966 to 1967.

The 1963 borings had shown that the area has a very complex geology. To explore the conditions of the soil, which consisted of sand and marly and clayey layers, and to extract undisturbed soil samples, 35 borings with a small diameter and 16 borings with a large diameter were sunk, and nine wells (2.5 × 2.5 m, depth—10 m) excavated. As they were also to serve observation purposes, the borings with the small diameter contained a plastic tube with a lean concrete skin (Fig. 6-19); the larger borings and the wells contained concrete tubes

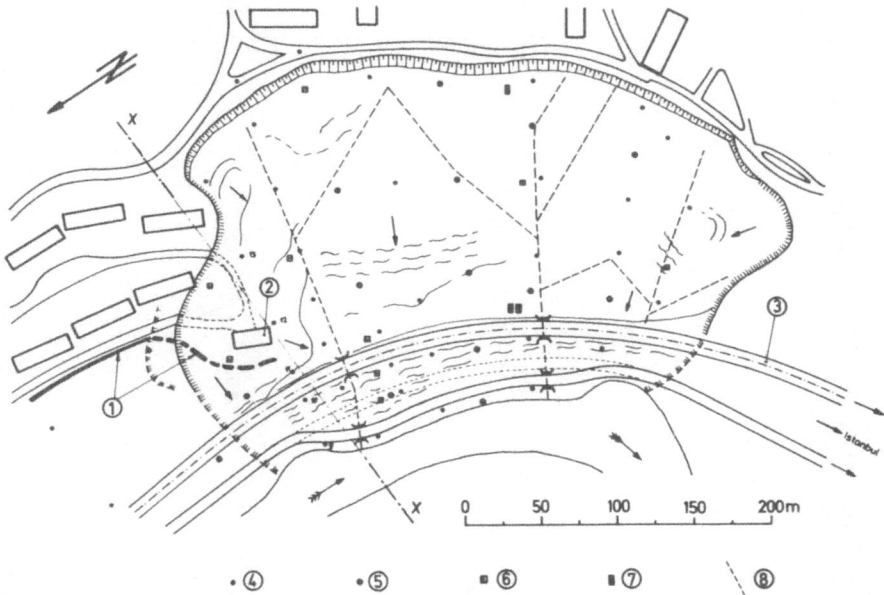

Figure 6-18 Site plan of landslide; (1) retaining wall; (2) sludge digestion tank; (3) railway tracks; (4) borings with small diameter; (5) borings with large diameter; (6) wells—2.5 × 2.5 m; (7) trial-loading of piles; (8) old drainage trenches.

with 20-cm and 60-cm diameters, respectively; thus 60 observation points were available to monitor soil movements. The borings containing concrete tubes served at the same time for the observation of ground-water conditions. The positions of the borings are shown in Fig. 6-18.

Ground-water conditions were quite irregular; for example, no water was found in some sand layers, while at other borings water was seeping from sand veins and from fissures in soft to very stiff marly clay. This might be explained by a very complex stratification of the layers that defies calculations and was created by the various previous landslides running in different directions, which had frequently changed the flow of ground water and in some places interrupted it. Of course the rising ground-water pressure, especially during the rainy seasons, together with the disturbed course of the stratification, was one of the causes for the slide movements.

Types of Soil Movement. From observations and measurements it was determined that these were composite slides on slip surfaces positioned between 7 and 10 m below ground surface. Since the slides were also triggered by earlier slides, the problem was that of successive slides. Figure 6-20 shows a section of the slope after the last slides in 1967, which dislocated the sludge-digestion tank and destroyed the retaining wall and the road in this area (see also Fig. 6-18). At the scarp the soil consisted of folded clay layers and limey material subjacent to stratified, fissured, clayey limestone.

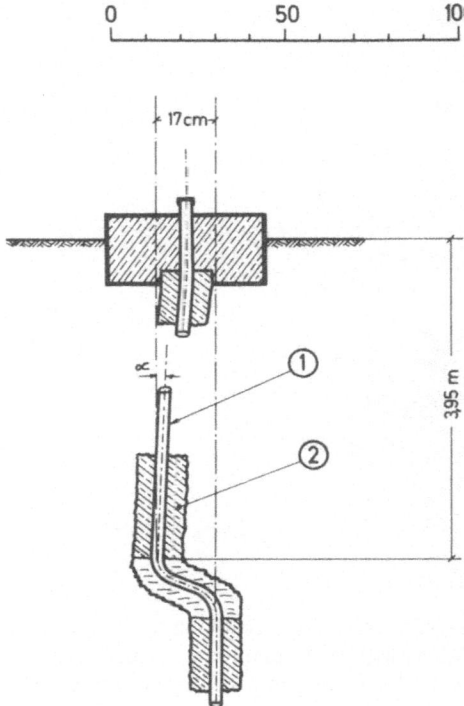

Figure 6-19 Dislocation of boring B11 (see also Fig. 6-20); (1) plastic pipe; (2) skin of lean concrete.

Stabilization Measures. Since an effective drainage of the whole slide area had to be considered virtually impossible, other alternatives had to be found to retain and stabilize the slope.

The laterally unconfined strength q_u, determined in the laboratory, varied between 25 kN/m² and over 400 kN/m², and this sometimes even with samples

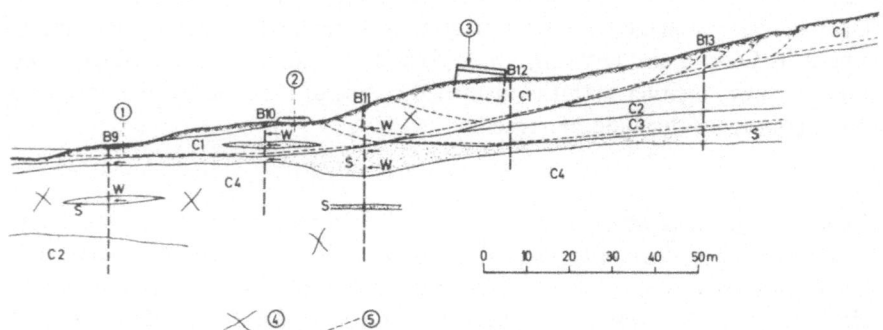

Figure 6-20 Section X – – – X of slide shown in Fig. 6-18; (1) road; (2) railway tracks; (3) sludge digestion tank; (4) cracks and occasional slickensides; (5) slip surfaces, S—sand, C1—yellow clay, C2—dark-green clay, C3—stiff, blue clay, C4—very stiff, green-to-blue clay, W—seepage water, B11—boring with small diameter.

from the same borehole. Shear tests that were repeated no less than 10 times yielded an effective shear angle φ' from 3° to 14° and an effective cohesion c' of 28 to 20 kN/m². As an approximate value for the shear stress in a recent slide (Fig. 6-20), τ was determined to have a magnitude of 18.5 kN/m². Now, in order to arrive at a safety factor of approximately 1.35, the total retaining strength required by a 50-m-long and 7-m-deep moving soil mass was estimated to be $\tau = 20$kN/m² and density $\gamma = 2.0$ t/m³. Since the soil consisted at greater depths of very stiff to hard fissured marl and marly clay, stabilization with bore piles was recommended. These piles were concreted after the contractor method. In order to test the effectiveness of the bore piles, horizontal trial loadings were carried out in three places on piles with a 60-cm diameter.

The following plan of action (Fig. 6-21), consisting of three phases, was set up to stop soil movements completely:

Phase 1: Construction of a retaining wall with bore-pile footings along the railway tracks (E–E), and of drainage trenches A, B, and D, as shown in Fig. 6-21.
Phase 2: Construction of drainage trench C.
Phase 3. Construction of a retaining wall below scarp (F–F).

The piles serving as footing for the retaining wall along the railway route have a 1.0-m diameter. The drainage trenches are 1.0 m wide and 7.0 m deep, for the backfill a material similar to that encountered in the sandy layers of the area was used. Long concrete girders with a 1.0-×-1.0-m profile were installed

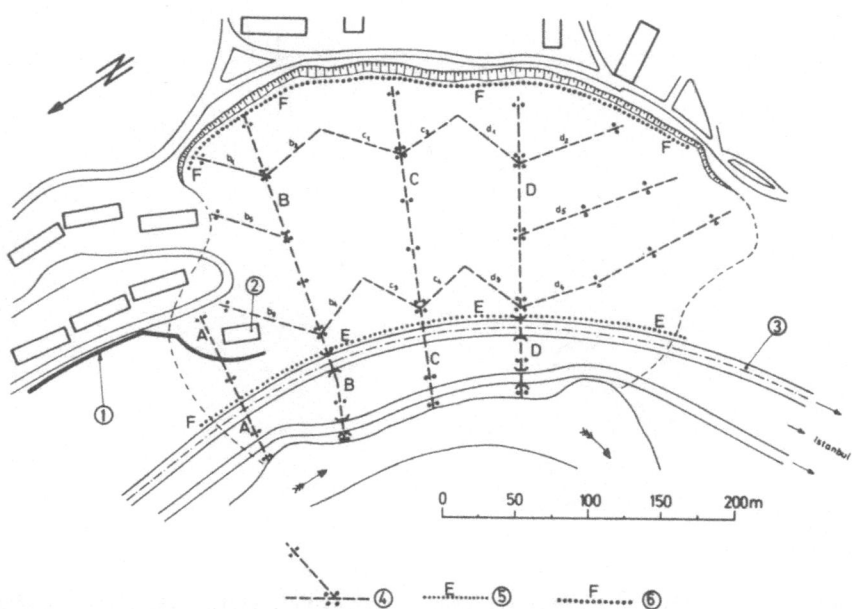

Figure 6-21 Site plan with completed stabilization measures; (1) retaining wall; (2) sludge digestion tank; (3) railway route; (4) drainage trenches with nailing piles; (5) retaining wall following railway line; (6) retaining wall at toe of scarp.

in the compacted granular material, serving to take up horizontal forces acting parallel to the ground surface. These long blocks act like hard cores that transmit the forces by friction to the nailing piles (Fig. 6-22).

The main points to be observed in carrying out this project were

- To avoid any additional constriction of the ground-water flow
- To avoid an increase of the pore-water pressure
- To stop all soil movements until drainage takes effect
- To prevent destruction of the drainage trenches by unexpected soil movements

Continual observations were made for indications that additional stabilization measures would have to be taken.

It was very important to carry out the plan of action consistently, step by step. The installation of slightly inclined drainage pipes in the sand layer (see 6.1.4.2 (I)) might have further improved ground-water conditions. The effectiveness of nailing piles is confirmed by Case History 6.1.2 (II).

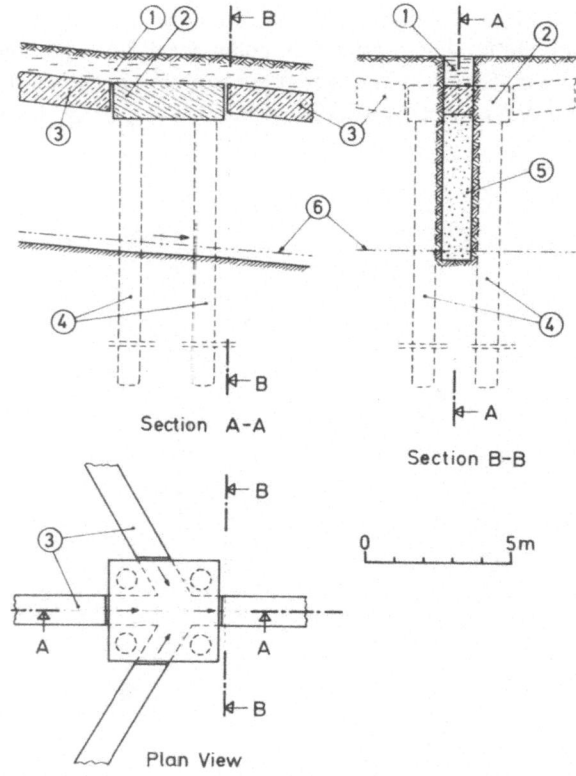

Figure 6-22 Details of drainage trenches branching out, with nailing piles; (1) clay covering; (2) head slab on piles; (3) concrete girder; (4) nailing piles; (5) drainage trenches; (6) slip surface.

6.1.2 (II)   Stahlberg Freeway, German Federal Republic (Sommer, 1977)
This is an example of a creep in stiff clay. In the course of freeway construction
in the German Federal Republic, a 12-m-high embankment had to be filled on
a slope with a gradient between 6° and 9°. When the fill reached a height of
10 m, the slope started to move. First, a ballasting fill carried out as an extension of the downhill embankment shoulder was attempted in order to stop the
slide (Fig. 6-23). Movements were measured continuously; they decreased
from 13 to 8 mm per month but it certainly could not be said that they had
been arrested completely. No doubt, this was a typical example of a creep
movement.

With an inclinometer it was determined that the position of the slip surface
was 15 m below ground surface. A recalculation of stability conditions yielded
a residual shear angle of 8° to 9°. The subsoil consisted of overconsolidated
tertiary clay; Figure 6-24 shows the soil coefficients. This tertiary clay was
highly plastic; below the slip surface its plasticity increased with the higher
clay content and the plasticity index and water content were higher and the
wet unit weight lower. The undrained shear strength c_u of 200 kN/m² below
the slip surface was twice as high as that above the slip surface. As described
repeatedly by Skempton, this was a slide problem not to be dealt with on the
basis of soil coefficients obtained in laboratory tests but a tectonic problem that
had been created by previous deformations of the slope; these had reduced the
original shear strength to its residual value.

The stabilization was carried out with reinforced-concrete wells with a 3.0-
m diameter, installed at center distances of 9.0 m. The wells reached 5 m into
the firm subsoil below the slip surface; their static effect may be compared to
that of bolting. Experiences gained at numerous slide stabilizations have

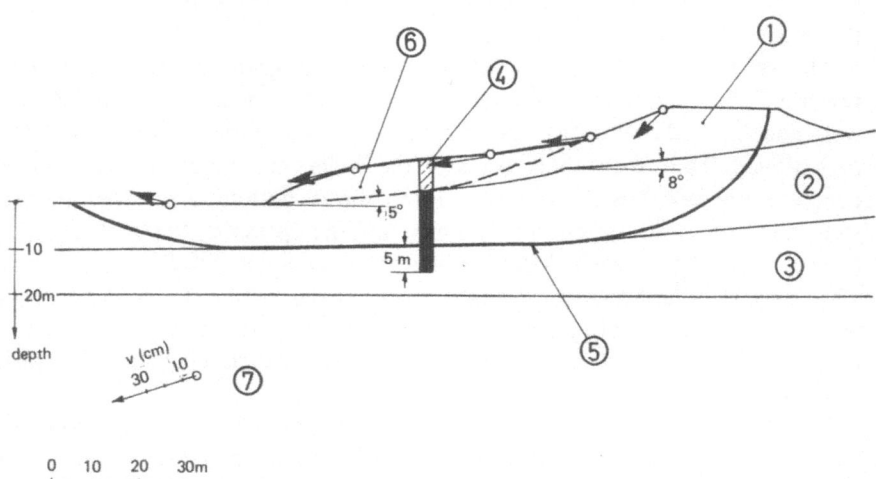

Figure 6-23 Section of slide area; (1) dislocated part of embankment; (2) moving
clay mass; (3) undisturbed clay; (4) reinforced-concrete wells; (5) slip surface; (6)
downhill embankment; (7) movement vectors.

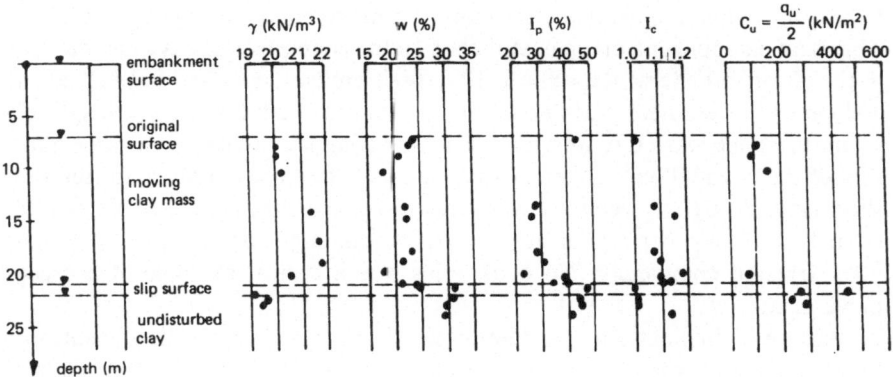

Figure 6-24 Soil-mechanical coefficients.

proved that only a minor increase of the safety factor from 1.0 to 1.05 or 1.1 may suffice to stop slide movements. In this case, a safety factor of 1.1 was used in the calculations, yielding a required resisting force of 0.8 MN/m². The dimensions of the wells and the distance between them were determined after Brinch Hansen, as a function of soil resistance and its distribution along the well shaft. Also of importance is the determination of soil resistance in the nonsliding area of the slope. After Brinch Hansen this is to be stated as the 7.5-fold value of the undrained shear strength c_u. The latter was determined experimentally to be 0.15 MN/m², yielding a maximum soil resistance of 7.5 × 0.15 = 1.125 MN/m²; therefore, the dimensions of the reinforced concrete profiles ensured an adequate stability. If creep movements continued, additional wells were to be installed. Owing to a gradual stabilization of the slope, this measure has so far not become necessary.

The wells (Fig. 6-25), 29 in total, were sunk with grabs and their walls were reworked with hammers. At the so-called slip joints, smooth, shiny slickensides were encountered over a distance of 2.0 m; in this zone the soil was completely mylonitized. The 10-cm-thick walls of the wells were made of cast-in-place concrete, reinforced with steel nets. From top to bottom, 2-m-wide, circular steel nets were inserted into the well hole and then pressed back with concrete (size of grain: 0 to 8 mm). After the walls were completed to their final depth, 20-m-long steel reinforcements built in two sections were installed. Concreting work was carried out after the Contractor method. Under certain conditions it may be feasible to use T-shaped diaphragm-well panels in place of circular wells (See 6.2.9.2).

In order to determine the effective earth pressures and thereby also to verify the static formula, earth pressure and pore-water pressure gauges were installed in several wells at exposed spots of the wall (Fig. 6-26). The measured stresses amounted to only 30% of the calculated values. As far as the figures as such were concerned, the calculation was not confirmed; however, the con-

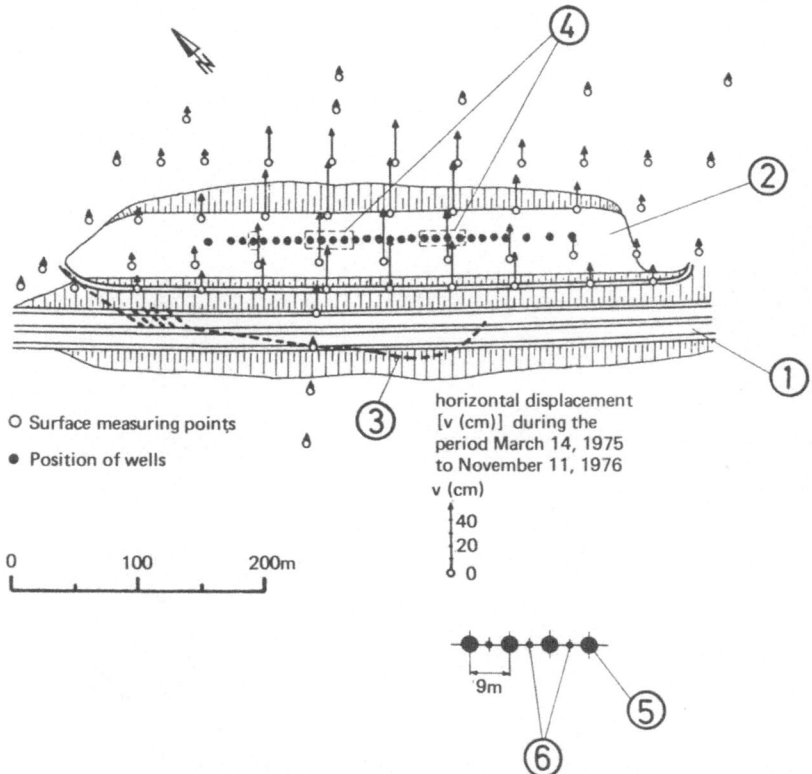

Figure 6-25 Arrangement of wells, with movement vectors; (1) route of motorway; (2) downhill embankment; (3) scarps of slide; (4) position of measuring wells; (5) measuring wells; (6) inclinometers.

centration of the measured earth pressure values in the driven slope immediately above the slip surface, as well as in the slope taking up the earth pressures below the slip surface, corresponded to the statement of Brinch Hansen. In addition to the earth pressures, pore-water pressure was also measured; it reached about 60% of total stresses. Furthermore, a number of surface and subsurface gauges were installed primarily to obtain indications of whether the soil between the wells would move more forcefully than the wells themselves. At the start of the measurements, displacements took place at the slip surface that increased towards ground surface. The mean displacement of the top rims of the wells amounted to 2 cm, resulting in a tilting of the well shafts by 1:1,300 (horizontal to vertical). Measurements at the slide joints and the slip surface indicated that the movement was gradually stopping. A decreasing slope pressure also pointed towards a stabilization of the slide. Holtz and Massarsch (1976) describe a similar stabilization by piles and Datye (1980) recommends the use of flexible piles to arrest slides.

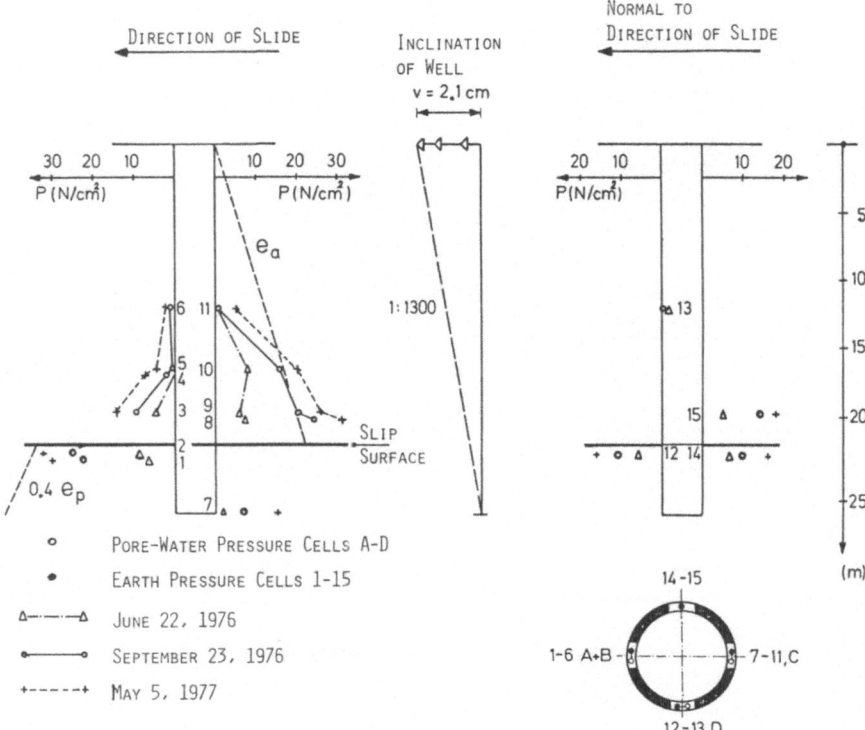

Figure 6-26   Measurements of earth and pore-water pressures.

6.1.3   Stabilizing Moving Structures: Stabilization of an Abutment for a Freeway Overpass near Graz, Austria

In early October 1967, after the cut for the route had been carried out, slide phenomena were observed around the uphill abutment of object G-19, an overpass for the southern freeway east of Graz, Austria. Slide movements were at first rather close to the surface, taking place in part between an upper layer of brown loam (sandy-clayey silt) and a subjacent layer of grey clayey silt, both originating in the Pannon. In some places a heavily water-bearing layer of sand and gravel was encountered below the clayey silt.

A few days later the slide assumed major proportions and a slickenside could be recognized near the abutment. The whole area around the abutment was affected by the slide and parts of the abutment were bared.

Since the landslide was evidently triggered by excavation operations for the freeway route on a slope that was near limit equilibrium in its original state, and since the slide was about to assume major proportions, an immediate stabilization of the uphill abutment was absolutely necessary.

Stabilization Measures
Diaphragm-wall panels, running parallel to the slide movement, were installed at the side and directly in front of the downhill face of the abutment (Figs. 6-27 and 6-28), to prevent the expected downhill displacement. These panels were interconnected at the tops to form a rigid system (Fig. 6-27) and fixed into the stiff *Opok* (local term for clayey silt) at the bottom. The horizontal forces generated by movements of the uphill masses sliding down on the slickesides were taken up by the diaphragm-wall panels and transmitted to the firm, stationary subsoil. Diaphragm-wall panels were chosen because of the low clearance under the bridge, which made the construction of bore-pile panels impossible. Another argument in favor of the diaphragm wall was the consideration that the slope must not be disturbed further so as not to trigger another slide when the stabilization measures were being carried out.

Dimensions of the Diaphragm Walls
The diaphragm walls were designed after the "Palisade"-wall theory (Fröhlich, 1959). Since the panels were horizontally firmly interconnected, a uniform deformation was assumed. The soil above the slickensides had an angle of friction of $\varphi' = 24°$ (disturbed soil sample, repeated shearing), but on account of the ongoing slide only $\varphi' = 12°$ was taken into account. The *Opok* had an angle of friction $\varphi' = 22.5°$ and a cohesion $c' = 0.09 \text{ MN/m}^2$. In the following, the rough calculations are shown:

Effective Forces

(a) Ea = Effective earth pressure, calculated for double width of abutment (affected zone)

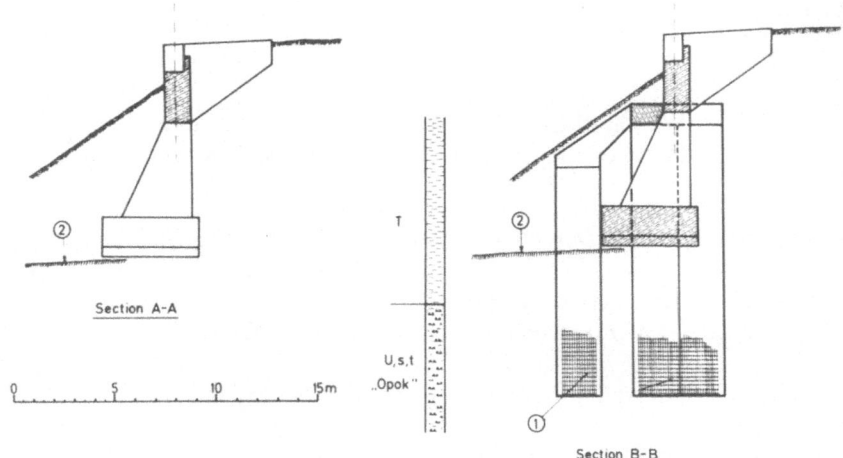

Figure 6-27 Sections of abutment, before and after stabilization; (1) diaphragm-wall panels; (2) slickensides. (For explanation of symbols used in soil profile, see 6.6.)

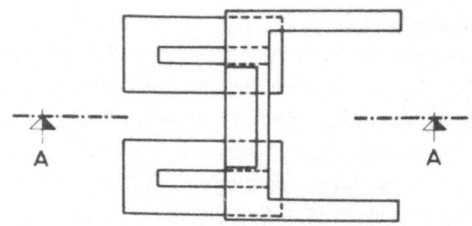

Plan View before Stabilization

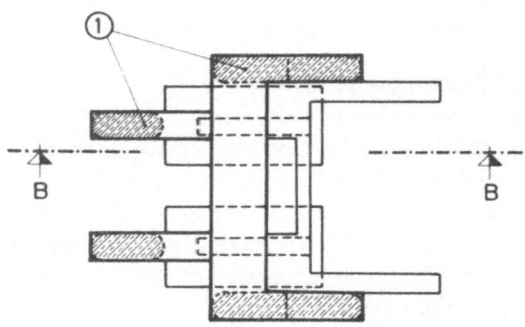

Plan View after Stabilization

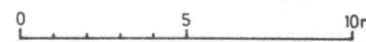

Figure 6-28 Plan of abutment, before and after stabilization; (1) diaphragm-wall panels.

(b) G = Weight of soil mass plus dead weight of bridge
(c) R = Force of friction on slickenside due to G, $R = G \cdot {}_{+g} \overline{\varphi}'$

Load Acting upon the Diaphragm Wall at the Level of the Slickensides

$M = Ea - R$ as horizontal force
M = Moment of Ea in relation to level of slickensides.

The diaphragm-wall panels take up this load by way of:

(a) Passive earth pressure (E_p) on front of panels
(b) Skin friction at sides of panels
(c) Bottom pressure of panels

Required Proofs. (1) Proof of the depth to which the panels need to be fixed into the *Opok:*

h = Difference in height between the slickensides and the bottom end of the diaphragm-wall panels. Reference point is the moment equilibrium around the center of the bottom:

M_r = Moment to be taken up by skin friction

M_E = Moment due to E_p and E_a, acting on the diaphragm walls (below the slickensides)

M_σ = Moment to be taken up by bottom pressure

$$\text{present } \eta = \frac{M_r + M_E + M_\sigma}{H \cdot h + M} \geqslant \qquad \text{required } \eta = 1.5$$

It will be expedient to investigate first whether M_r and M_E alone might be capable of providing the required safety. If this is not the case, then the residual moment ΔM has to be transmitted to the bottom joints

$$\Delta M = \text{required } \eta(H \cdot h + M) - M_r - M_E.$$

(2) Proof of safety against mechanical ground failure:

According to well-known practice (Terzaghi and Peck, 1967), a required safety factor $\eta = 1.5$ is to be assumed.

The decision to use diaphragm-wall panels offered the advantages of stabilizing the existing structure without damaging it, avoiding additional bracing, and not further disturbing a slope already near limit equilibrium.

Other stabilization methods, such as the construction of an underpinning from an open braced pit or the construction of foundation wells, would have caused a relaxation of the soil even if carried out with utmost care; this would have led to dangerous soil movements and serious damage to the structure that was already in a labile state.

6.1.4 Stabilization by Reduction of Pore-Water Pressure

6.1.4.1 Reduction of Pore-Water Pressure with Drainage Trenches

6.1.4.1 (I) Stabilization of a Slope near Retznei, Styria, Austria. Owing to heavy rainfall during June and July of 1972, landslides occurred near Retznei which seriously threatened the railway line between Graz and Spielfeld, a part of the vital Austrian north–south connection. The slide area extended over approximately 100 m.

The formation of the ground surface is shown in Fig. 6-29. Following the line of the dip, it was divided in an upper zone, Slope I, a zone of transition, Slope II, and a berm next to the railway tracks.

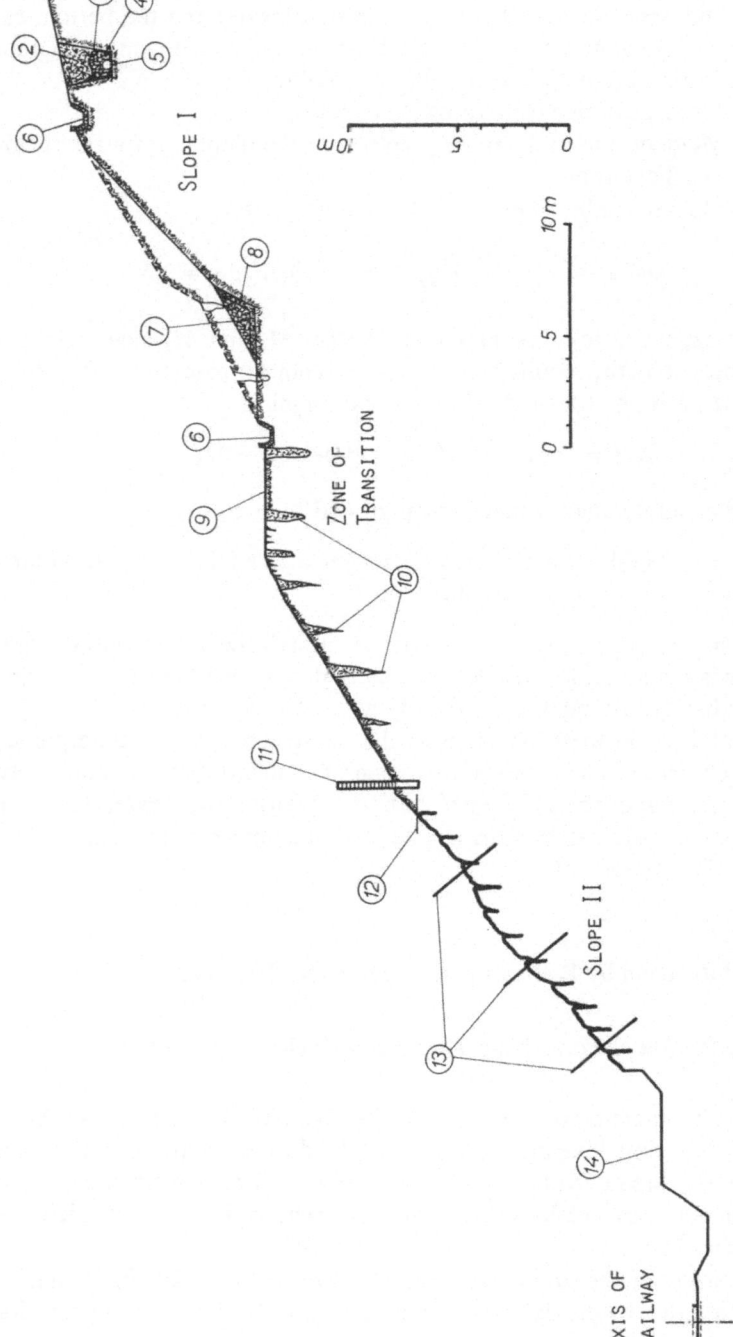

Figure 6-29 Section of slide area; (1) orchard on crown of slide; (2) backfill in trench; (3) filter fabric; (4) gravel; (5) drainage pipe; (6) open drainage; (7) large boulders (toe wedge); (8) layer of concrete gravel (0–30 mm), 10 cm thick; (9) path; (10) sand mixed with cement, water, and bentonite; (11) wall constructed of rails and sleepers; (12) bedrock surface; (12) horizontally tensed catch-nets; (14) berm.

The upper zone already showed signs of slide movements in the form of folded-up heavings. The soil in this zone consisted of a 2-to-3-m deep layer of loose, weathered material on top of partially weathered sandstone beds; it had a mean gradient of approximately 10°. In Slope I below, which had a gradient of about 35° and was densely overgrown with trees and bushes, evidently a number of slides had taken place. Some of the scarps were several meters long and in places the displaced material had a pulpy consistency.

In the so-called zone of transition there were open cracks originating from slides which reached 2 to 3 m down to the sandstone bed below; here and there unweathered sandstone was cropping out.

Slope II, with a gradient of 40°, mainly consisted of unweathered exposed sandstone. Blocks of weathered marly soil had descended and collected at the foot of the slope and on the berm.

The sandstone had a multilayered stratification, with the dip running parallel to the ground surface. It had to be feared that future heavy rainfalls would loosen more soil blocks that would move in the direction of the railway tracks.

Stabilization measures. In order to achieve an effective collection of surface waters in the upper zone that normally flowed down to the railway, a surface drainage was carried out consisting of an open channel running fairly parallel to the contour lines of the sandstone bed; to prevent an overflow, the downhill rim of the channel was provided with a lip.

Uphill from the open channel a 2.0-m-deep trench was excavated to achieve a subsurface drainage. A slotted plastic tube (20 cm in diameter) was placed in the trench, covered with coarse gravel and wrapped in filter fabric (see also 6.1.4.1 (II)). Inspection shafts were installed at distances of 50 m to monitor the effectiveness of the drainage trenches. Thus the whole water of the upper zone was collected and diverted to a main drainage before it could reach Slope I.

The uppermost loose soil blocks on Slope I were removed and the slope secured with a toe wedge. This wedge, serving as a filter, consisted of riprap on a 10-cm-thick bed of aggregate. Now the surface water from Slope I could be collected in a surface drainage and diverted to the existing main drainage. The open cracks in the zone of transition were filled with a mixture of fine sand, cement, and bentonite.

Slope II was first cleared of all remaining soil blocks down to the sandstone bed. The blocks that had collected on the berm were removed. The head of Slope II was secured by a fence constructed from railroad rails and sleepers. The rails were rammed to a sufficient depth (approx. 0.8 m) into the firm soil and cemented. In order to intercept any sliding soil blocks, catch nets, consisting of steel rods driven in at right angles to the ground surface, and nylon mesh, were strung over the whole slope. The steel rods were placed at distances of approximately 5.0 m vertically and 4.0 m horizontally. To avoid an overloading of the catch nets, the accumulated soil is cleared away every 2 to 3 years. The stabilization measures rendered the relocation of the railway route unnecessary and have so far proved effective.

As with almost any landslide, the following three actions were of prime importance:

(1)   Diversion of surface waters
(2)   Closing of major cracks
(3)   Sowing grass on the slope surface

In addition, the following measures were required in this case:

(4)   Subsurface draining of the crown of the slide
(5)   Construction of a toe wedge for the upper slope
(6)   Securing of the lower slope with a combined rail-sleeper fence and nylon catch nets

Other stabilization measures, and by far more expensive ones, such as shot-concrete, anchorings, or even retaining walls, had become superfluous.

6.1.4.1 (II)   Landslide Graz-Ruckerlberg, Austria. On the northern slope of the Ruckerlberg, which is from the Pannon and is not very steep, a landslide occurred that represents a type of slide that is relatively frequent. The affected area was approximately 130 m long, with a maximum width of approximately 100 m; the average gradient was in the range of 9° to 11°. The eastern part of the slide area was used for cattle grazing and farther to the west were residential buildings and orchards. Between pastures and orchards was a graben which served as natural catchment for surface waters. The ground surface was marked by numerous cracks and bulges; there were also a number of small ponds which held water permanently, and in many places the topsoil was marshy (Fig. 6-30).

From 1966 through 1969 a two-storey residential building with basement had been constructed on the western part of the crown above the slide area. In the course of construction work a 3.5-m-high fill, supported by a retaining wall, was carried out.

In order to explore the soil structure, a shaft was sunk in the uphill zone of the slide area and undisturbed soil samples were examined in the laboratory. Three borings, sunk at different spots, completed the picture (Fig. 6-31). The surficial layer, reaching to a depth of 2.5 m, consisted of firmly deposited silty sand with a permeability coefficient $k_f = 10^{-7}$ cm/sec. This layer had such a high silt component that it was almost impermeable and acted as a water barrier. The subjacent layer, reaching to a depth of 5.0 m, was of loosely to medium-densely deposited sand with a permeability coefficient $k_f = 10^{-3}$ to 10^{-4} cm/sec. This layer was heavily water-bearing; the ground-water table was encountered at a depth of 3.15 m.

Below a depth of 5.0 m there were alternating layers of sand and silt and at a depth of approximately 8.0 m followed stiff, impermeable, clayey silt (*Opok*). These layers extended over long stretches nearly parallel to the slope surface; in some places the water-bearing sand layer was wedging out. When the inflow of water to the sand layer increased, for instance during heavy rainfalls, then a point would be reached where more water entered the layer than could flow

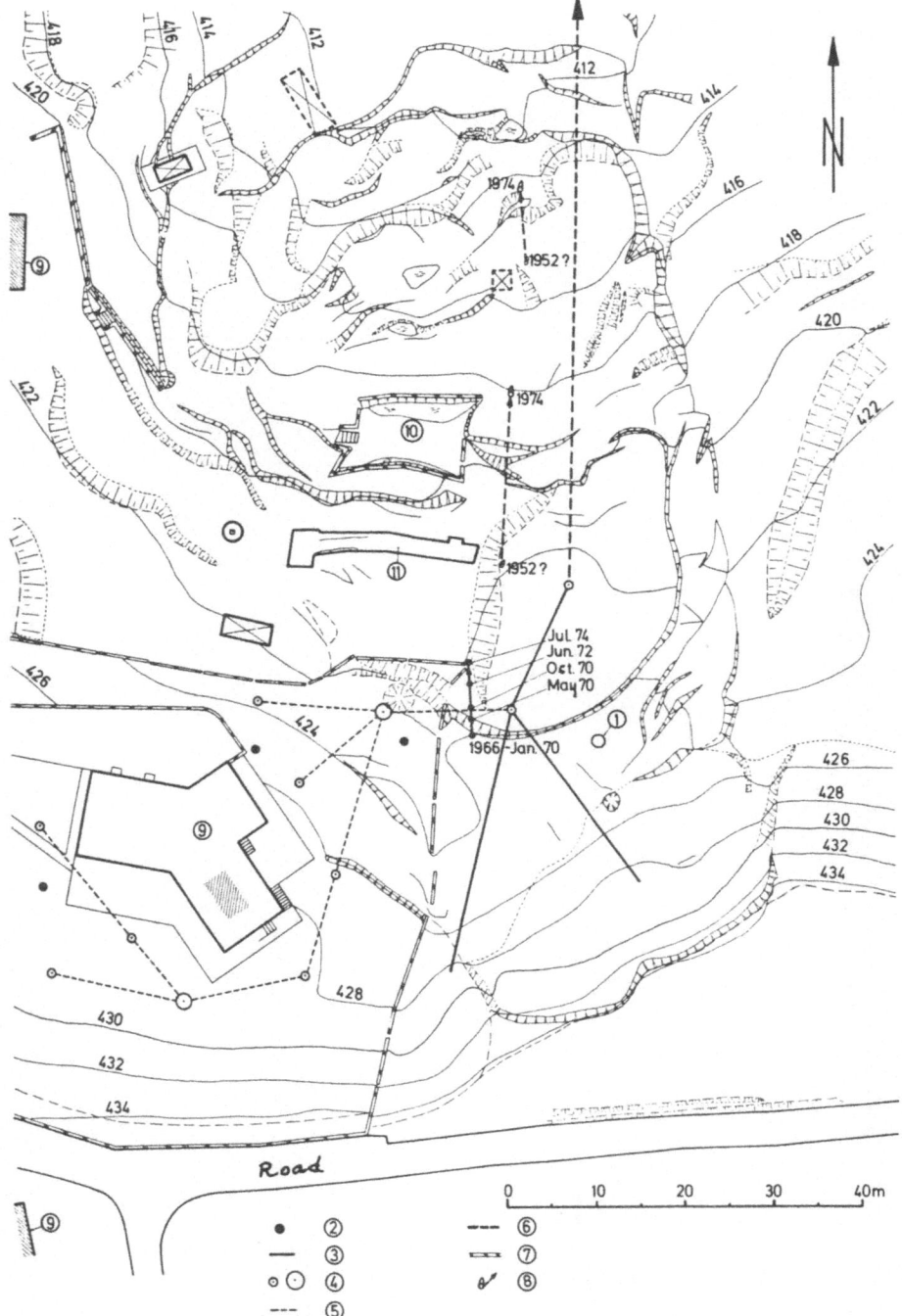

Figure 6-30 Site plan of slide area; (1) test shaft; (2) test borings; (3) drainage with filter fabric; (4) seepage wells—diameter 1.0 and 2.0 m; (5) pipes connecting seepage wells; (6) diversion of water through plastic pipes; (7) fence; (8) dislocation of boundary stones; (9) building; (10) swimming pool; (11) bowling alley. (Geodetic survey: Rinner, Graz, Austria)

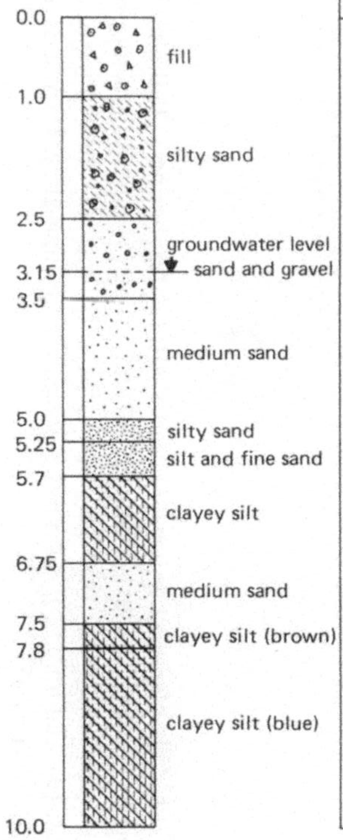

Number of sample	Depth m	e —	m_v m²/kN	k cm/sec	T_{LABOR} sec	T_{NATURE} day	C' kN/m²	φ' °	W_n %	Unit dry weight loose γ_d kN/m³	Unit dry weight dense	Unit dry weight natural	d_{60} mm	d_{10} mm	d_{60}/d_{10} —
3443	1.0-1.5	0.60	$1.9\cdot10^{-4}$	1.10^{-7}*	720	47	20	28	17.8						
3447	2.1-2.3	—	—	—	—	—	10	30	22.7						
3448	2.3-2.55	0.626	$1.4\cdot10^{-4}$	$1.4\cdot10^{-7}$*	400	26	20	31	20.5						
3449	2.55-2.8	—	—	—	—	—	—	—	16	14.19	—	18.8			
3452	3.2-3.45	—	—	—	—	—	—	—	16.2	14.04	—	brown 17.4 blue 19.8	1.4-1.6	0.05-0.08	18-30
3456	4.1-4.3	—	—	—	—	—	0	36	13.6						
3460	5.0-5.25	—	$\sim2.0\cdot10^{-4}$	5.10^{-5}+	10	—	0	36	12	—	—	silt 15.9 sand 16.2			
3461	5.25-5.45	0.618	$0.8\cdot10^{-4}$	8.10^{-8}*	400	—	60	22	26						
3463	5.5-5.7	—	—	—	—	—	55	13	32.3						

*) k-value found from Consolidation Test.
+) k-value found from Permeability Test.

0.0
1.0 — fill
2.5 — silty sand
3.15 — groundwater level
3.5 — sand and gravel
5.0 — medium sand
5.25 — silty sand
5.7 — silt and fine sand
6.75 — clayey silt
7.5 — medium sand
7.8 — clayey silt (brown)
10.0 — clayey silt (blue)

Figure 6-31 Soil-mechanical coefficients, landslide Ruckerlberg near Graz, Austria.

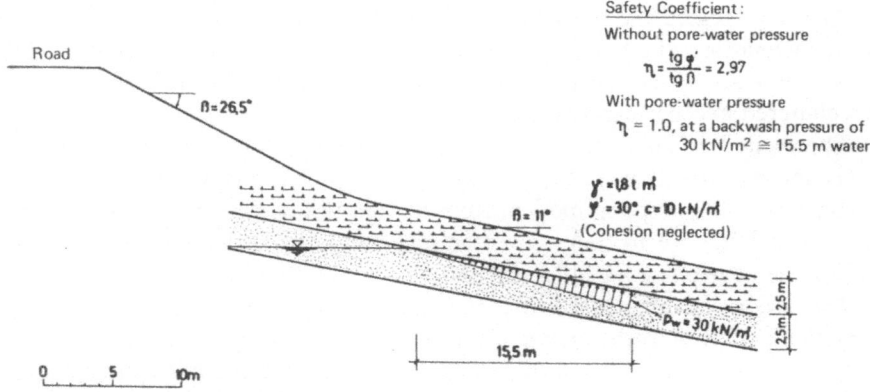

Figure 6-32 Section of slide with water backwash.

off, creating a backwash (Fig. 6-32). This rising of the ground-water level caused by the backwash could also be registered by measurements taken in the boreholes. Here the pore-water pressure acts against gravity and in the end the shear resistance may be smaller than the driving forces. This mechanism creates such strong surface movements that—for example—from 1970 through 1974 a fencepost moved 8.3 m (Fig. 6-30). In the process certain zones of the sliding layers underwent deformations as illustrated in Fig. 6-33. Here it should be noted that this profile corresponds exactly to the geodetic plan and has not been overstated.

Prior to the slide, the slope had a uniform gradient of 11°. By the action of the water pressure sections B_1 and F_1 were pushed upward by 3.0 m, separating these two originally connected zones. The steeper embankment E_1, with a 1:0.7 inclination (vertical to horizontal), and the embankment C_1, with 1:1 counter-

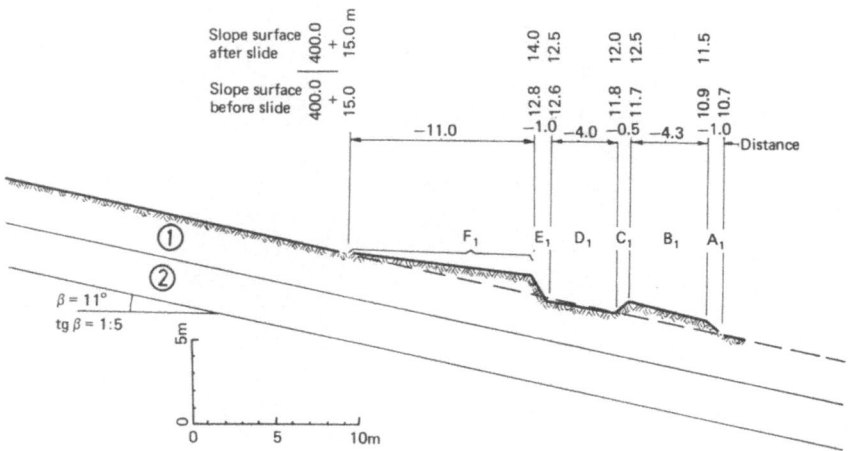

Figure 6-33 Deformation of sliding layer.

inclination, formed the graben D_1; a bulge advanced from section B_1 with a 1:1.2 inclination. The bulge A_1, consisting of material already loosened, had a lower inclination than sections E_1 and C_1. This surface formation was also encountered in other parts of the slide area.

The whole landslide consisted of several individual slides. Stated briefly, the mechanism of the slides was the following: the action of the water pressure was setting the whole slope in motion, from bottom to top, whereby the preceding slide would remove to quite an extent the support for the zone above, thus aiding the development of the next slide.

The serious question was whether there was any connection between construction activities at the crown of the slope and landslides, especially in view of fact that they had only started in 1969, that is, after the completion of the residential building. This building, plus the fill, represented a mean additional stress of 24 kN/m². Laboratory tests showed that under such a load the subsoil would be fully consolidated after about 50 days at the most, but the landslides occurred from 1970 through 1972, starting almost 1 year after the completion of the building. The foundation for the building reached slightly into the water-bearing sand and this did cause a minor constriction of laminary water flow in the sand layer; however, this change in the water flow was so minor that an effect on the slide movements could be excluded.

Some damage was evident at the building itself, such as the development of cracks between house and ground and some cracks from heavings in the basement due to the weak design of the floor (Fig. 6-34); but it appeared certain

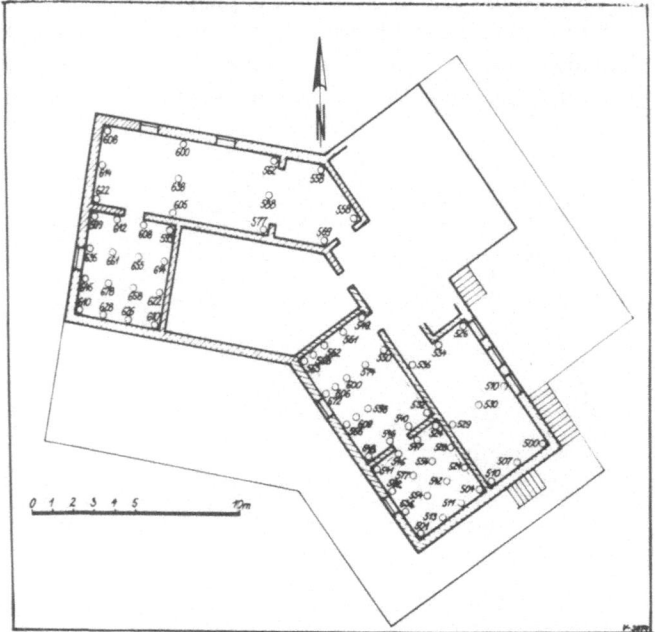

Figure 6-34 Uplift of underdimensioned basement floor, height notations in millimeters.

that the construction of the building and the fill had not caused or triggered the landslides.

Stabilization measures had to aim primarily at dealing with the high water pressure in the sand layer. The original plan called for draining the slope with subsurface drainage, to be installed uphill and downhill of the residential building. Due to difficulties with the neighbors and concerns regarding the construction of subsurface drainage close to the building, this plan was discarded. A combination of drainage wells and subsurface drainage (Fig. 6-30) was then carried out.

The wells were positioned at the sides and downhill of the building (8 wells, diameters—1.0 and 2.0 m, respectively), reaching to depths between 4.5 and 5.5 m, depending upon the position of the water-bearing sand layer. The water was diverted by horizontal or slightly inclined pipes connecting the wells.

The subsurface drainage was installed at the head of the main slide some distance from the side of the house. As in the case of the wells, the depth of the trenches, with a total length of about 70 m, was chosen on the basis of the subsoil structure, reaching down by over 4.5 m. The trenches were carried out as filter fabric drainage (Fig. 6-35); this ensured lasting effectiveness. The water collected in the wells and drainage trenches was diverted to an inspection shaft and from there by a PVC sewer pipe (diameter 125 mm) to the main drainage (Ragnitz Creek). This combination of wells and trenches completely eliminated the high water pressure in the vicinity of the endangered building. The initial water discharge of over 1 liter/sec decreased shortly to a fraction of this flow.

In this case, the water-bearing sand layer was of such a dimension that other measures, for example the installation of horizontal drainage pipes, would not have been effective. Secondly, the existing building had to be taken into consideration, and it was necessary to keep the ground-water table at a constant level in its vicinity, regardless of seasonal fluctuations of water supply from rainfall and melting. As it was, there was no need to touch the building or to construct an underpinning, e.g., with bore piles. By the measures taken, the area around the building was drained so efficiently that all slide movements stopped.

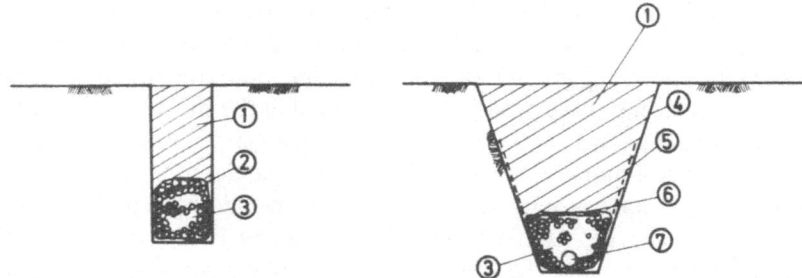

Figure 6-35 Two types of deep drainage; (1) fill of low permeability; (2) filter-fabric wrapping; (3) gravel; (4) side of trench; (5) filter fabric while gravel is filled in; (6) wrapped filter fabric; (7) perforated plastic pipe.

6.1.4.1 (III)   Landslide Kleinsölk, Styria, Austria. Starting at the capture, the
alignment of the channel supplying the Sölk power plant with water from the
Kleinsölk Creek was designed to cross, with a length of approximately 500 m,
an area with a latent tendency to slide. In spring and summer of 1977 construction work for a road that was to follow the channel on the uphill side triggered
quickly progressing ground failures in the zone of the future channel and in
the uphill zone; up to 0.5-m-wide cracks opened and the scarps were up to 1.5
m high.

On the uphill side of the channel alignment the slope had predominantly an
inclination between 1:4 and 1:1.25 (vertical to horizontal). Further uphill, at
a distance of 50 to 80 m, the slope leveled off to 1:6, followed by another steep
zone. The major part of the topsoil had a markedly marshy character.

Where the channel was to cross the slope, the soil mainly consisted of slightly
clayey, finely sanded silt with thin continuous sand layers as well as occasional
silty talus, stemming from the rock above. The rock consisted of quarzite mica
containing hornblende, with varying shares of mica, hornblende, and garnet.
Soil explorations in the main body of the landslide showed basically the same
soil structure. The cohesive layers were distinctly plastic and as water-saturated as the intermediate sand layers. Where the slope leveled off, soil explorations encountered talus containing loam (sandy-clayey silt) to varying
degrees, with a marked water flow. The discharge into the trenches was estimated to amount to 5 to 10 liters/sec and in the case of a strong backwash the
ground-water table would rise to the level of the surface. Prevailing soil conditions and the resulting action of the ground water were the causes for the
landslides. The silty, finely-sanded soil had a permeability coefficient that was
several orders of magnitude below that of the uphill talus, causing in the tran-
sitory zone a backwash from the continuous and strong inflow of ground water;
this created a corresponding water pressure. There were also a few springs.

In this case, the only technically and economically acceptable solution was
to drain the slope by subsurface drainages, and these were carried out during
the winter of 1977–1978 (Fritsch and Prodinger, 1980). A total of more than
1,500 m of drainage trenches were constructed and provided with filter fabric
drainage (Fig. 6-36; see also 6.1.4.1 (II), Fig. 6-35), reaching to depths of over
5.5 m. The individual drainage trenches reached well into the uphill talus zone,
whereby the major draining effect for the slope was achieved in that area and
the backwash pressure lowered. The drainage trenches farther downhill,
branching out laterally, mainly intersected the mass banking up the water, and
collected the remaining ground water and seepage water; they aimed especially
at draining the continuous sand layers embedded in the silt. By staggering the
trenches in this manner, a reduction of the backwash pressure and a quickly
progressing consolidation of the cohesive soil were achieved.

The drainge trenches extended through a 100-m-wide zone uphill from the
500-m-long slide area. At the points where drainage lines joined, inspection
shafts were installed and the discharge of the individual trenches was measured. For the trenches reaching into the talus zone the maximum value was
more than 5 liters/sec (January 1978), while the cohesive material naturally

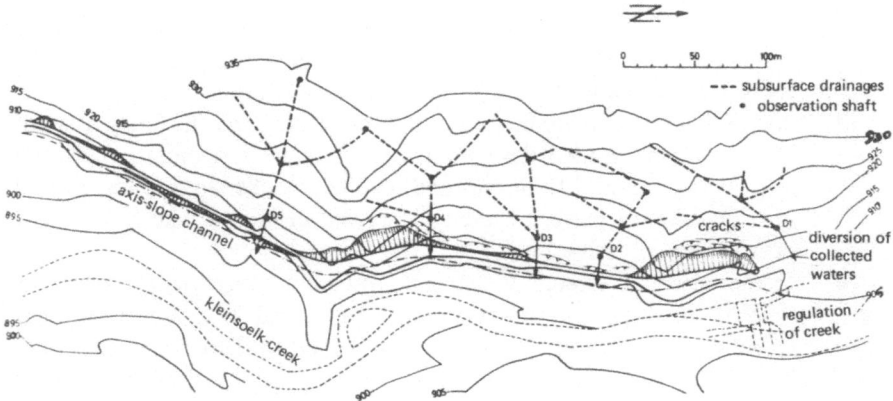

Figure 6-36 Site plan of slope drainage, Kleinsölk.

yielded much lower quantities of water. In the course of construction work there were evident signs of soil consolidation and by the time the drainage system was completed the slope movements had stopped altogether.

6.1.4.1 (IV) Budapest, Dunaújváros, Hungary (Kézdi, 1976). The 1964 landslide extended by 1,300 m along the approximately 50-m-high precipice following the bank of the Danube. A pumping station situated near the riverbank was shifted 35 m towards the Danube. On the terrace extending from the precipice lies the town Dunaújváros.

The subgrade below the precipice consisted of silty sand (natural water content $w = 0.20$; plasticity index $w_p = 0.18$; liquidity index $w_L = 0.49$) with occasional intermediate clay layers ($w = 0.20$; $w_p = 0.25$; $w_L = 0.49$). Geologically this was a winddrift of Pliocene loess, having a higher vertical than horizontal permeability because dead roots from plants had created vertical cavities. Subjacent to the approximately 22-m-deep mass of loess were Pannonic sediments consisting of clay and layers of fine sand. At a depth of about 15 m there was an upper free ground-water table; a lower, artesian ground-water table was rising from a depth of 30 m almost to the ground surface.

The landslide was triggered by a rising of both ground-water tables by approximately 12 m that was caused by the construction of industrial facilities on the terrace (seepage). The slip surface was at a depth of approximately 40 m below ground surface in a horizontal, silty-clayey layer.

Stabilization measures
(1) Construction of a 5.5-km-long jetty, with transverse dams at fairly short intervals, to prevent an erosion of the slope toe. The core of the jetty was filled with cheap blast furnace slag; the embankment facing the river was covered with riprap, the side facing inland with a layer of gravel.
(2) The squares formed by the transverse dams were filled one-third with gravel and two-thirds with loess; the latter was obtained when the

ledges were cut into the precipice. The sandy gravel was filled so as to form ribs at a right angle to the river in order to divert the water from the slope over its whole length without obstructions.

(3) Along the southern, down-stream stretch of the bank the finely fissured loess wall was graded down to 1:2 (vertical to horizontal); there the slope was covered with humus and planted. Wickerwork fences placed in a chessboard pattern served to prevent erosion of the humus and proved very effective.

(4) In the upstream stretch of the slope three cuts were made, each from 8 to 10 m deep, to tap the ground water and to divert it through trenches to the Danube; the cuts also served to increase evaporation. They were backfilled with crushed rock on top of a filter bed, which held back the fine particles while the ground water could flow off freely.

(5) Ledges 6 m high and 6 m wide were cut into the part of the precipice situated below the town; their horizontal planes were inclined towards the land, where open concrete channels diverted the water from the slope. In addition, vertical cuts were made to dry out the soil.

(6) A system of drainage ducts 709 and 400 m long was installed in order to lower the ground-water table below the surface and to bring down the artesian water in the loess hill behind the pumping station.

(7) In addition to the ducts, 26 wells were bored at distances of approximately 16 m, and 20 observation wells were installed.

(8) At the head of the loess slope 34-m-deep shafts with an inner diameter of 2.8 m were sunk.

In the loess zone the shafts were lined with steel tubing to a depth of 21.0 m and with brickwork in the subjacent silt, behind which a gravel filter was placed. In order to extend their radius of action, horizontal wells were advanced at a depth of 30 m (see 6.1.4.3). From the bottom of the 34-m-deep shafts, filter wells were bored to a depth of 43.5 m as a safeguard against uplift. The combined effects of these measures led to a stabilization of the slope.

6.1.4.2 Reduction of Pore-Water Pressure by Horizontal Borings from the Ground Surface

6.1.4.2 (I) Landslide Graz-Ries, Austria. During the night of July 15, 1972, a landslide occurred on the Ries, situated east of Graz in an area originating in the Pannon, which was quite characteristic for that part of the country. Heavy rains had preceded the slide, with a total precipitation of 159.7 mm within 6 days.

The affected area extended with a length of approximately 85 m parallel to a road. Over a length of approximately 75 m it dipped in a southwesterly direction; the part to the northeast had a gradient of 33% for a distance of 30 m, while the lower, southwestern part had a gradient of only 11%.

On the crown of the landslide were several buildings: Block A (15 × 12 m)—a two-story residential building with strip foundations resting on timber piles (diameter—20 cm) spaced about 1.5 m apart, reaching depths of 4 to 5 m; Block B (12 × 11 m)—a single-story indoor swimming pool with a 50-cm-thick, strongly reinforced concrete slab (no timber piling) as a foundation; Block C (17 × 12 m)—residential building on strip foundations with timber piles (similar to Block A) (Fig. 6-37).

Description of the ground surface. At the head of the slide a marked, almost 2-m-high scarp had developed, touching tangentially the southern front of Block C, passing under Block B, and bypassing Block A at a distance of 2 to 3 m. Consequently, the safety of all three blocks was threatened and Block C already showed some cracks.

Downhill from Block A was a graben (slump) with steep uphill and downhill scarps. At the foot of the slide was a bulge with numerous rolls and ponds. This

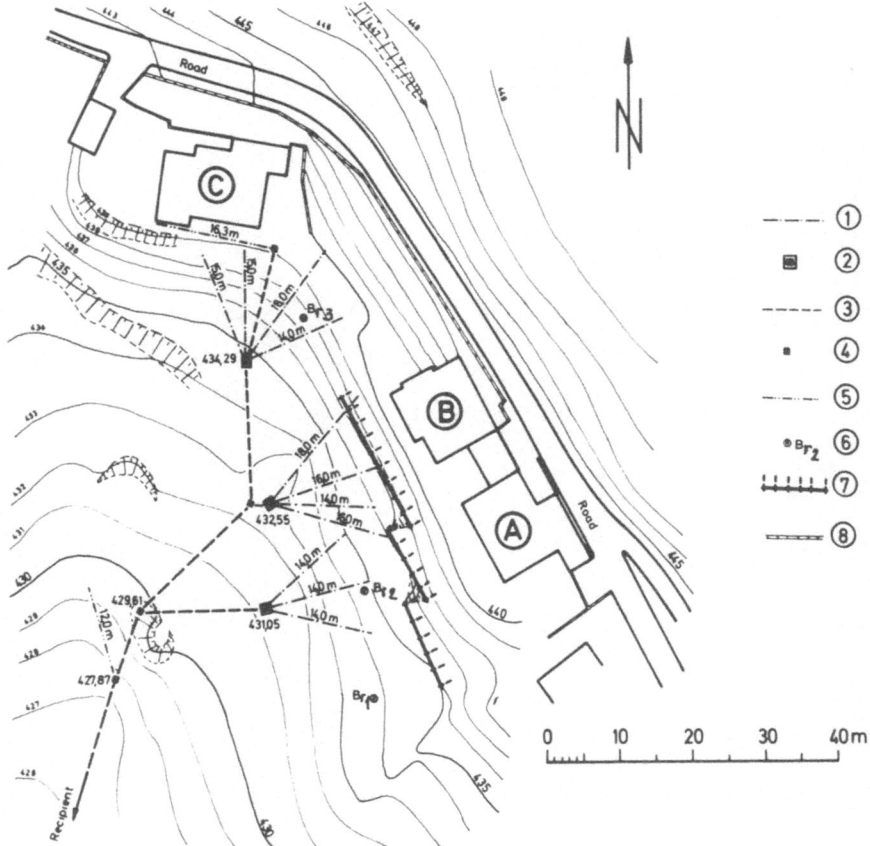

Figure 6-37 Site plan of slide area with completed stabilization measures; (1) horizontal filter pipes, diameter—4 cm; (2) construction pits and shafts, 80 × 80 cm, for the horizontal drainage pipes; (3) connecting pipes, diameter—8 cm; (4) collecting shafts, 50 × 50 cm; (5) drainage pipes, diameter—8 cm, in trench; (6) wells, (7) "Krainer" wall; (8) retaining wall.

evidence led to the conclusion that, starting in the area of the graben, soft soil layers were being pressed out (Fig. 6-38). In the slide area trees were tilted and later on died.

The flanks of the slide and the zone uphill from the road were marked by irregular bulges from earlier slides and by tilted trees. In the moist zone below the swimming pool (Block C) an old ineffective drainage was discovered during the excavation of a trench.

Description of subsoil. To explore subsoil conditions and to investigate the causes for the landslide, a shaft (inner diameter—1.0 m) was sunk precisely in the very steep downhill scarp of the graben to obtain two soil profiles at the same time (Fig. 6-38). The soil profile for the uphill side showed that from a depth of about 4.5 m the soil was loosely deposited and contained embedded

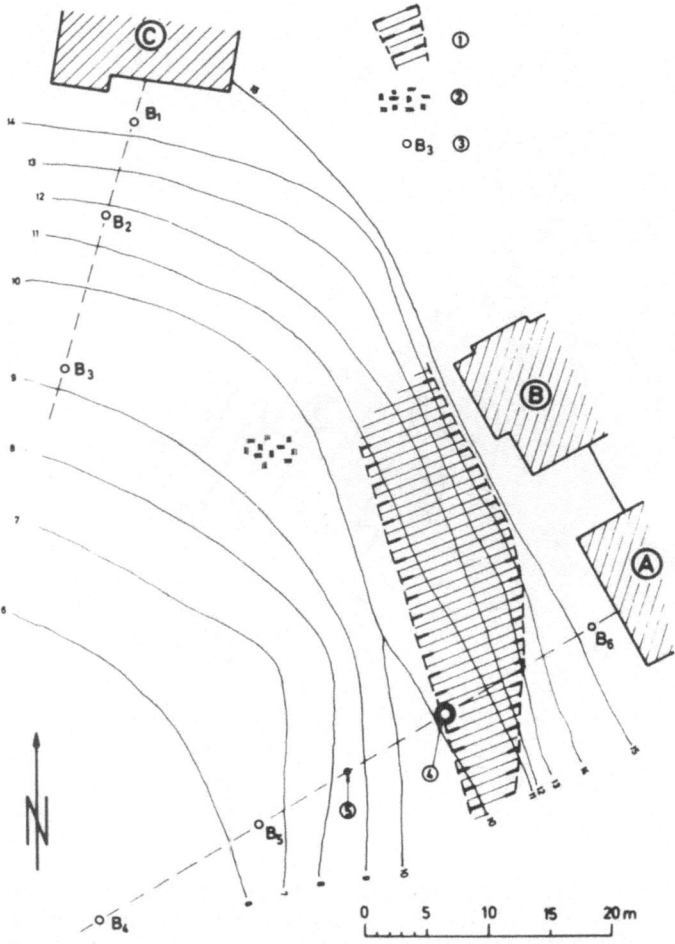

Figure 6-38 Site plan of zone mostly affected by slide; (1) slumped part of terrain; (2) moist area; (3) boring; (4) shaft; (5) piezometric tube.

sand lenses; only from a depth of 6.5 m did it become more dense and loamy (sandy-clayey silt). The downhill soil profile was more or less similar, but the soil had a higher density. *

A sample extracted at a depth of 5.0 m showed the following grain distribution:

0 —0 .002 mm (clay)	approx.	1%
0.002—0 .06 mm (silt)	approx.	44%
0.06 —2 .0 mm (sand)	approx.	31%
> 2 .0 mm (gravel)	approx.	24%

In addition, six exploratory borings were carried out to further investigate the soil and to gain a picture of water conditions. The position of the water tables in the different borings and the shaft (Fig. 6-39) as well as the loose consistency of the water-bearing layer confirmed the assumption that the increased horizontal water pressure and the increased pore-water pressure in general, and the softening of the soil due to water absorption in particular, were the causes of the landslide.

In the shaft, water was rising relatively quickly and stabilized at 2.6 m below ground surface. There was the imminent danger of renewed slide movements as soon as rainfall set in again.

Stabilization Measures. In order to stabilize the terrain, the causes for the slides had to be eliminated, i.e., measures had to be taken to reduce water pressure in the soil. From soil explorations it was evident that the water-bearing layer was about 6 m below ground surface; therefore, an effective drainage trench would have to reach at least to that depth. In view of local conditions, a technically sound construction of such deep trenches would have been extremely expensive.

By a pumping test carried out in the shaft it was investigated whether it would be possible to lower the ground-water table with wells. A piezometric tube was placed between the shaft and Boring B5 (Fig. 6-38) to monitor the ground-water table. A volume of 75 to 80 (100) liters/hr was pumped out, but due to the low k_f-value of the soil the area of effectiveness was limited and consequently it appeared doubtful whether a lowering of the ground-water table by wells would be successful.

The following stabilization measures were carried out: First, three wells (diameter—32 cm; filter tubes' diameter—12.5 cm; depth—6.0 m) were bored. These wells were used for further soil explorations and for extended test pumping, which partially drained the quicksand and facilitated the installation of horizontal filters in the quicksand to the required lengths.

Three pits (2.0 × 3.0 m; depth—2 to 3 m) were excavated and temporarily braced. From these pits three to four horizontal borings (lengths—14 to 18 m; diameter—12.5 cm; cocos-filter tubes, diameter—4.0 cm) were made in a fan-like pattern. These horizontal drainages were joined in shafts (80 × 80 cm), which were later concreted, and the collected water was diverted with PVC

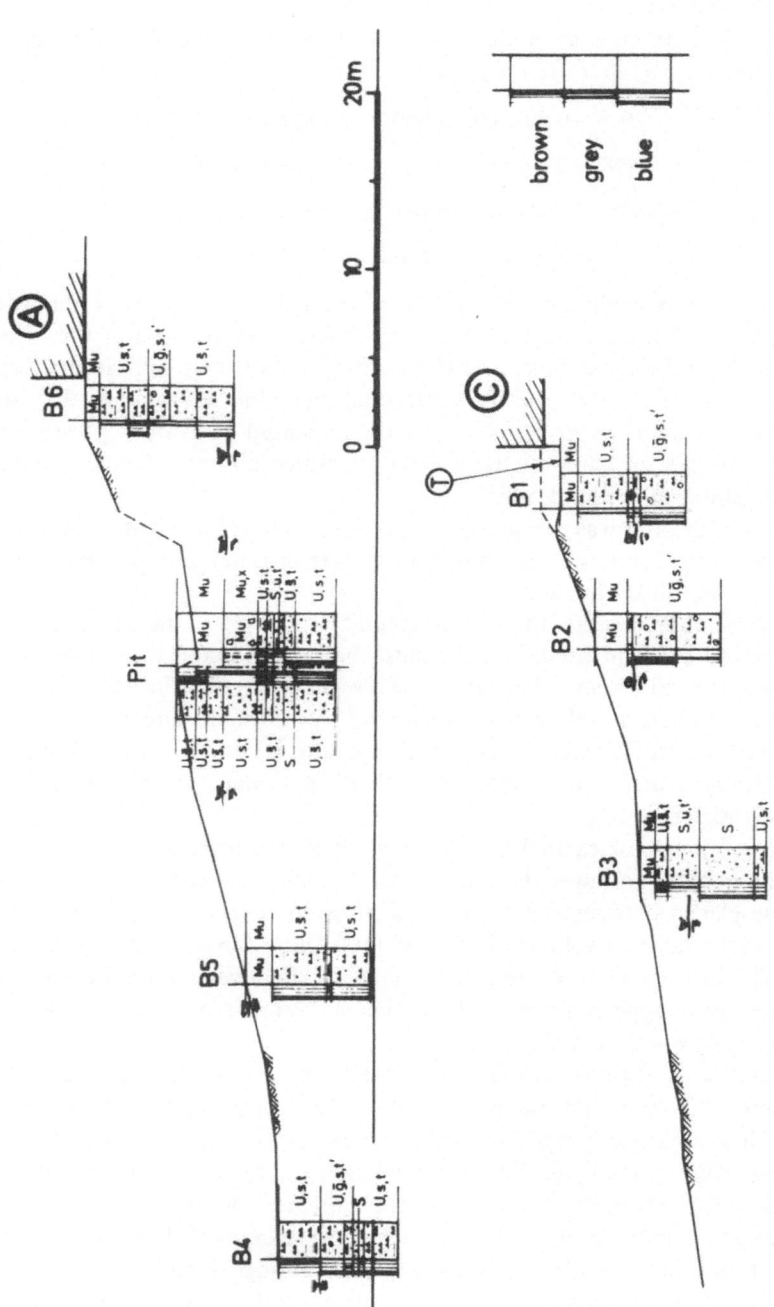

Figure 6-39 Section of slide area, with soil profiles; T—slumped terrace; B—borings. (For explanation of symbols used in soil profiles, see 6.6.)

pipes (diameter—8.0 cm). The arrangement of drainages and shafts is shown in Fig. 6-37. By mid-1976 the rate of water discharge from the horizontal filter tubes was about 0.3 liters/sec.

Next, drainage trenches were installed directly south of Block C and in the downhill part of the slope. PVC-drainage pipes were placed into the 40-cm-wide and, on the average, 1.5-m-deep trenches and surrounded by a gravel filter; then the trenches were backfilled with excavation material. The soil left over from the excavations was used to fill existing cracks. Furthermore, wooden "Krainer Walls" (walls constructed of wooden or concrete beams, placed one upon the other at uniform distances, vertically and parallel to the slope—6.2.9.5. (I)) were constructed in the zone southwest of Blocks A and B. The downhill foundation strip of Block C was underpinned with four wells (diameter—1.0 m).

To complete the stabilization effort, the whole area was graded and the slopes were compacted, covered with humus, and seeded in order to prevent seepage of surface waters as much as possible. Since 1973 the whole slide area has been completely stable.

The measures taken eliminated the cause of the landslides, namely, the very high pore-water pressure during and áfter rainfalls, by relatively simple and economical means.

The construction of deep-foundation underpinnings for all three blocks, which was considered originally and is an extremely expensive solution, was therefore not necessary.

6.1.4.2 (II) Memphis, Tennessee, United States. Another example of the damaging action of water flowing in sand layers is the slump in Memphis, Tennessee, decribed by Redlich *et al.* (1929) and Terzaghi (1931), that occurred in July 1927.

At Memphis, the Mississippi River flows under a precipice about 30 m high and eventually cut into a sand layer cropping out below water level. This layer was under the pressure of á 10-m-high water column and had been prevented from flowing out by a huge superimposed clay layer. Where the clay layer was eroded by the river, the ground-water flow set the sand in motion and this in turn caused the whole sand layer to flow out.

The carbonization plant, situated 30 m above the river level at a distance of about 70 m from the river bank, together with a strip of land 200 m long and 30 m wide, slumped with a velocity of about 30 cm/hr to a depth of 20 m, tilted, and moved by about 15 m towards the river. The sand had an effective grain size of about 0.2 mm.

A timely prevention of this disaster would have required extensive drainage with horizontal filter tubes and the securing of the toe with Terzaghi filters.

6.1.4.3   Reduction of Pore-Water Pressure Using Wells with Horizontal Drainage—Landslide Kirchschlag, Lower Austria

In March 1970 an embankment slide badly damaged Federal Highway Nr. 61 from km 0.9 to 1.03 on the stretch between Kirchschlag and Karl (Fig. 6-40).

Figure 6-40 Motorway embankment after landslide.

The road embankment, from the downhill toe to the crown, had a maximum height of 12 m; the crown was 8 m and the roadway 7 m wide. Both flanks had a 1:2 inclination (vertical to horizontal).

The embankment was crossing an originally marshy depression; its base course had an inclination of approximately 1:4 to the southwest. During construction work for this road, which replaced an old road with numerous curves situated further uphill, landslides had occurred in the zone between the old and new road. At that time drainage trenches and a horizontal filter well were dug on the downhill side of the old road. This horizontal filter well, from which a PVC-pipe (diameter—10 cm) led to Shaft 1 (Fig. 6-41), proved very effective. The water collected there, along with surface waters, was diverted through a heavy-duty culvert passing under the new embankment.

The material obtained from a cut carried out for an adjacent road section in the direction of Kirchschlag was used as embankment fill. In 1968, during embankment construction, isolated initial cracks developed and the embankment profile was graded down because the fill material was not very suitable.

In March 1970, after heavy melting, the slide referred to at the beginning of this section started. Diagonal scarps developed in the road paving and the embankment body. The culvert passing under the embankment body was broken by soil movements and this led to further soaking of the subsoil.

Survey of the Ground Surface. On the uphill side of the embankment graben-like depressions running parallel to the roadway were visible, which marked the borderline between moving and stable soil. On the uphill side of the embankment small scarps shaped like seashells developed, which were also characteristic of the slopes of cuts made in this area. The embankment body was bulging out on the downhill flank. There the embankment material was loose, fully water-saturated and of pulpy consistency, similar to the material to be found in the bulges at the toe of landslides. Later on, numerous sliding blocks formed in the zone downhill of the embankment, where an irregular slanting of the embankment was also noted.

These observations led to the conclusion that the slip surface had mainly developed within the embankment body and did not reach far into the natural soil. This assumption was confirmed by the fact that trees were tilting quite irregularly. The heavy water saturation of the, as such, well-compacted embankment fill was caused by the unusually high inflow of water from the melting of snow. Contributing factors were the high longitudinal gradient and the one-sided transverse inclination of the roadway; the fill represented a large catchment area for precipitation and by necessity all surface water flowed into the loose downhill road shoulder.

Description of Subsoil. A number of exploratory borings were sunk and yielded the following soil structure (see Fig. 6-41):

Embankment—A sample from the embankment contained less than 3% clay, about 26% silt, and about 11% fine sand; it also contained coarse sand and stones. The natural angle of internal friction was above 30° and the soil also had, in a moist state, a small cohesion. If fully water-saturated, however, this cohesion was lost, and soils of such a composition mostly tend to flow. The base course of the embankment consisted in part of humus; roots were found in all borings.

Surficial layer of natural soil—This layer extended to depths of 4 to 5 m, was moist and plastic where the main component was silt and dry where the main component was sand. In this layer the soil was loose to medium-dense.

Subjacent layer—This layer had a much higher share of coarse sand and a lower share of silt. Gravel lenses and stones were frequently encountered; they cause, at the place of embedment, a discontinuity of the grain distribution in the silty coarse sand.

Stabilization Alternatives. (1) The first alternative taken into consideration was a *stabilization of the existing embankment*. This would have meant the stabilization of the slowly moving embankment and the elimination of the causes for the slide. The following measures were foreseen:

— Excavation of a 3-m-deep and 3-m-wide trench at the downhill toe of the embankment that would be backfilled with riprap. Installation of a perforated concrete pipe at the bottom of the trench, surrounded by filter gravel and wrapped in filter fabric, preferably on top of a lean concrete layer

— Loading of the embankment toe and flank with ballasting fill on top of a 30-to-40-cm-thick filter bed reaching into the embankment body. This would reduce the inclination to 1:2.5 (vertical to horizontal)

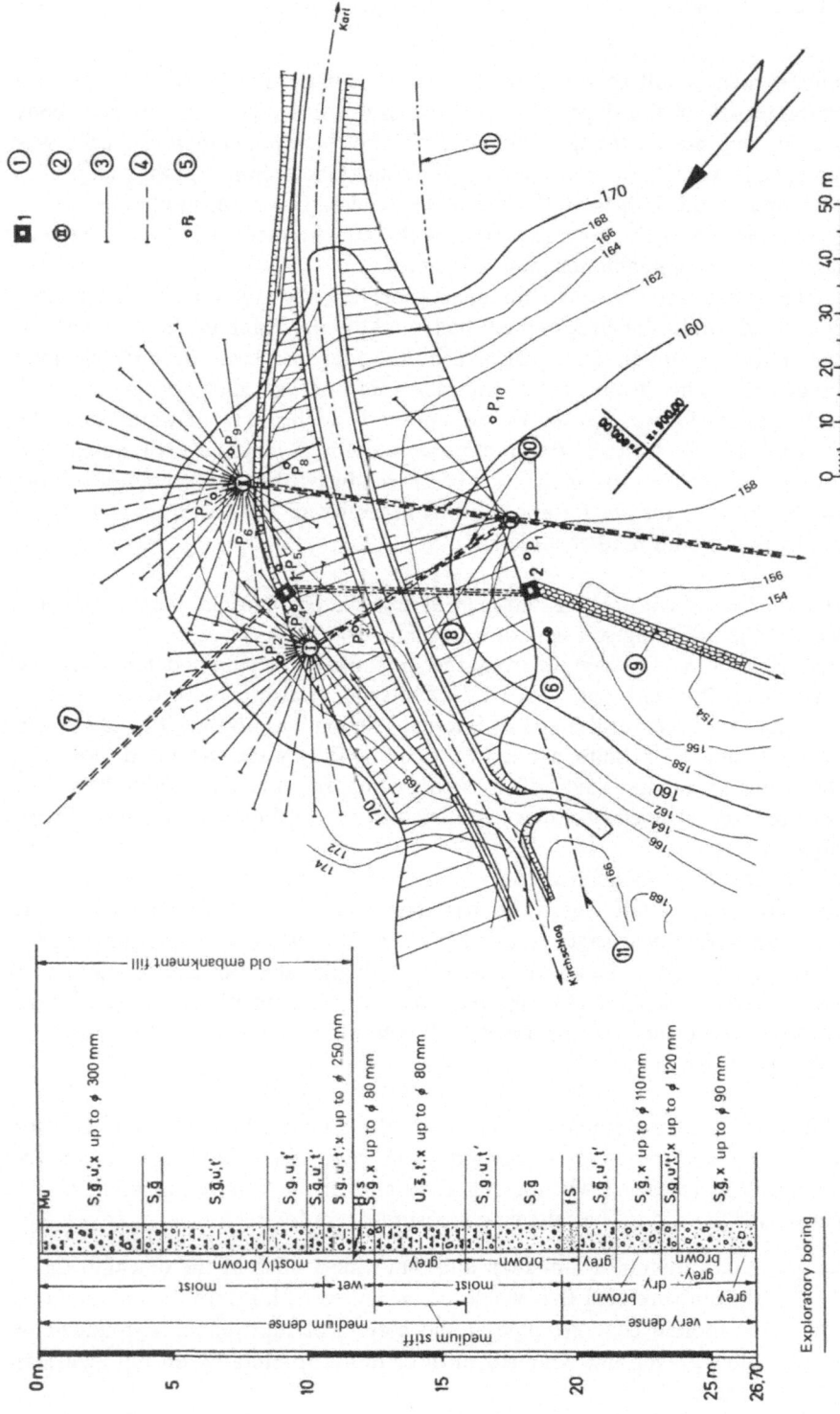

(Figure 6-41)

— Cutting-off the ravine with a 3-m-high and 8-to-10-m-long retaining wall and diversion of the water with a reinforced channel

— Construction of 1.5-to-2.0-m-deep seepage trenches and a collecting shaft at the uphill side of the embankment

— Construction of a new culvert that discharges near the downhill retaining wall

— Diversion of the water from the paving of the road by open channels at both sides of the roadway, and controlled diversion over the embankment flanks

— After an observation period of about 1 year, the road was to be paved.

(2) *Crossing the slide area with a pier bridge.* In view of the high costs of an embankment stabilization, the alternative of a pier bridge was investigated in August 1970. A three-span bridge (41–49–41 m) seemed to offer a viable solution; however, a cost comparison showed that this alternative would be 50% more expensive than the stabilization of the embankment. The advantage of the pier bridge was that it would have been an absolutely safe alternative.

The abutment on the Kirchschlag side was to have a spread foundation; the two piers and the abutment on the Karl side deep foundations. The measures required for this alternative were:

— Removal of the embankment to a position below the base course; grading down of the slope to 1:2.5

— Construction of pier footings with panels as slim as possible (diaphragm wall or overlapping bore piles), e.g., with a 0.6 to 0.8 × 6.0-m profile, which offers the smallest possible plane to the thrust of driving forces but still has a sufficient lateral bending resistance

— Diversion of the water from the zone uphill of the old embankment by an open cut.

Projecting work for this bridge lasted until the beginning of 1972.

(3) *Uphill-relocation of the embankment and stabilization of the subsoil by horizontal wells* (the alternative carried out—see Fig. 6-41). For cost reasons, in December 1972 the pier bridge alternative was finally dropped. Cracks that extended far beyond the sites foreseen for the abutments would have required the bridge to be longer than originally designed; extensive drainage would also have been required here. So the embankment was relocated uphill, thus reducing the length and height of the embankment and also avoiding the steep terrain to the southwest (creek ravine).

Figure 6-41 Site plan of slide area with completed stabilization measures and soil profile from one characteristic boring; (1) collecting shaft; (2) horizontal filter wells; (3) fan of horizontal filters 5 m below rim of well; (4) fan of horizontal filters 2 m below rim of well; (5) pore-pressure gauges; (6) exploratory boring; (7) plastic pipes, diameter—10 cm, leading to existing well; (8) heavy-duty pipes, diameter—80 cm; (9) paved channel; (10) galvanized-steel pipe, diameter—20 cm; axis of dislocated road.

In September 1972 five trenches were dug in the zone foreseen and it was found that the soil consisted of brown loam (sandy-clayey silt) to a depth of 0.7 m and from there to a depth of about 1.5 m of grey-blue silty loam containing an increasing share of sand with increasing depth. Below a depth of 1.5 m water was encountered everywhere.

Since high water inflow and the resulting increased hydrostatic pressure, and the flow pressure upon the slide body as well as the deterioration of the soil properties because of soaking, were the main causes for the embankment slide, the ground-water table was lowered with horizontal filter wells. Two wells (I and II) were excavated uphill and one well (III) downhill of the new road alignment. These wells (inner diameter—3.0 m; depth—approx. 8.0 m) were sunk by sections using the shot-concrete method. The horizontal filter drainages (plastic tubes wrapped in bast; inner diameter—4 cm, outer diameter—6 cm), extended by up to 30 m in a fanlike arrangement from the wells; uphill in two tiers, downhill in one tier. In part the borings were carried out as core borings. At the bottom of the wells a 30-cm-thick layer of filter gravel and a 50-cm-thick concrete base were installed, and a boring (diameter—15 cm; filter tube diameter—10 cm), to serve as relief against uplift, was advanced to a depth of 10 m below the well bottom. The water collected in the wells was diverted to the creek ravine by galvanized steel pipes (diameter—20 cm; Fig. 6-41). The total water discharge into the horizontal filter wells amounted, for example, on November 12, 1974, to 7 liters/min, and on July 1, 1975, during rainfall, to about 13 liters/min. The effectiveness of the filter tubes and diversion pipes must be checked twice a year. In order to monitor how effective this drainage was in lowering water pressure, pore-water pressures were measured with gauges (Glötzl system) before and after their installation. As the filter wells went into action, a large reduction of pore-water pressure was registered; it stabilized at a low value and only showed minor fluctuations from precipitation.

The uphill surface water was collected in an open channel and a foot drainage, diverted to Shaft 1, and from there by a heavy-duty pipe (diameter—80 cm) to the downhill Shaft 2. In addition, a 30-to-40-cm-thick filtering bed of gravel (0 to 30 mm) was spread over the whole base course of the embankment, but first the layer of brown loam had to be removed down to the surface of the grey-blue silt to eliminate a possible slip surface.

The embankment body was filled in layers 25 cm high that were individually compacted. The embankment flanks had a maximum inclination of 1:2.0 (vertical to horizontal). The remaining parts of the old embankment were graded down or removed. As a final stabilization measure, the area was seeded without spreading humus, and trees with deep roots were planted in the parts that previously carried trees. Since then, no further soil movements have occurred.

Of the three alternatives described above, the last was the one that tackled the evil at the roots—it reduced the damaging pore-water pressure. The first alternative would have stabilized only the embankment but not the subsoil, and on an unstable subsoil another embankment slide could have been triggered at any time by heavy precipitation. The second alternative, the construction of a bridge on piers founded on diaphragm-wall or bore-pile wall panels, could have been carried out quite quickly; however, it would have been much more expensive. Another alternative, the construction of a retaining wall at the foot of the embankment, was in practice not feasible without the drainage measures described above, and was therefore rejected.

6.1.4.4 Reduction of Pore-Water Pressure with Short-Circuit Conductors (after Veder)

6.1.4.4.1 Overview
by F. Hilbert

Stabilization with this method is achieved by ramming short-circuit conductors (e.g., steel rods or water pipes) vertically into the soil until they intersect the slip surface and reach 1 to 2 m into the subjacent homogeneous soil.

As may be learned from 7.2.2.4.2 and 7.2.2.5, this method is primarily effective for the stabilization of slides of sandy-clayey silt (loam) and clay soils, if a distinct slip surface exists between reducing (blue to grey) and oxidizing (yellow to reddish-brown) soils. In such cases the slip surface may be caused by electroosmotic water accumulation. Electroosmosis is only possible if there is an electric potential difference and a flow of current in the pore water; therefore, it can be eliminated by insertion of metal conductors that eliminate potential differences between soil layers. The electric current flowing through the short-circuit conductors is in principle generated by the reduction of components of the oxidizing soil and the oxidation of components of the reducing soil, or by the dissolution of the short-circuit conductors themselves (Veder, 1963). The process is shown schematically in Fig. 6-42. In these cases the slip surface normally follows the zone of contact between oxidizing and reducing soil, but is part of the reducing soil layer (see 7.2.2.5). Due to the ample presence of oxidizing agents that take up electrons (symbolized in Fig. 6-42 by iron (III) ions, but, e.g., oxygen from the air and manganese compounds may also take part in the process), electrons are transferred from the metal to the soil in (1). In the reducing soil layer (2) there is already a surplus of electrons, so that here the short-circuit conductor can only take up electrons, for example, by the reaction $Fe^{2+} \rightarrow Fe^{3+} + e^-$, the released electron being taken up by the metal. The dissolution reaction of the metal itself is equivalent, and leaves two electrons in the short-circuit conductor for each iron (II) ion generated. These electrons flow off in the short-circuit conductor and are released in the oxidizing layer by a reduction reaction. Basically, this means that positive charges are generated in the reducing layer while positive charges disappear in the oxidizing layer and the corresponding electrons migrate into the short-circuit conductors. This generation and disappearance of positive charges must be compensated by an electric current flowing in the electrolyte (pore water) in the opposite direction. As the negative charges in clay and loam are virtually immobile, current flow is only possible by migration of positive cations in the opposite direction. This flow of cations causes an electroosmotic transport of water in the same direction (see 7.2.2.4.2), which reduces the water content of the slip zone, as shown in Fig. 6-42 by the H_2O in brackets. It is favorable for

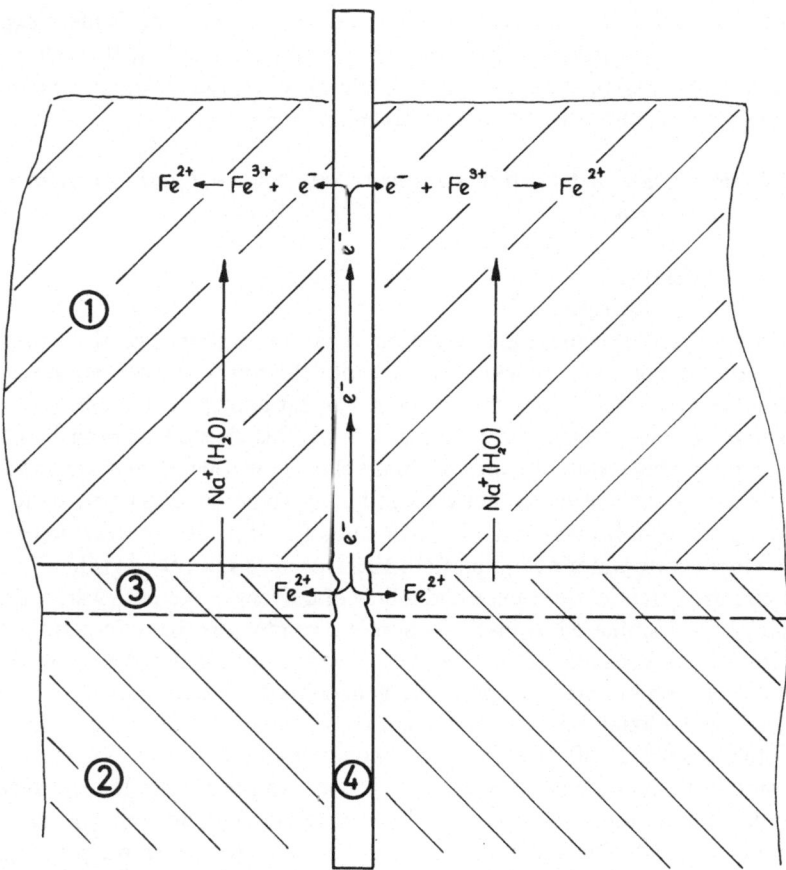

Figure 6-42 Model of the processes in a short-circuit conductor, after Veder; (1) oxidizing soil layer (e.g., brown sandy-clayey silt); (2) reducing soil layer (e.g., dark mudstone); (3) soft slip surface (high water content); (4) short-circuit conductor (steel rod). The metal of the short-circuit conductor is gradually attacked and after a number of years will be completely corroded, but the consolidation achieved by drainage and ion exchange is preserved.

the stabilization that the dissolution of the short-circuit conductors mainly takes place in the water-rich slip zone, since iron and aluminum will dissolve only if water is available to hydrate the emitted ions. Short-circuit conductors therefore have three main effects:

(1) They eliminate natural electroosmosis, which effects a flow of water to the zone of contact between oxidizing and reducing soils, by removing electric potential differences.

(2) Besides this, they cause an electroosmosis in the opposite direction which drains the slip zone.

(3) Finally, monovalent ions from the vicinity of the short-circuit conductors are carried off and replaced by higher-valency ions, which also causes a consolidation (7.2.2.3).

For practical application it should be noted that the short-circuit conductor method, after Veder, is only applicable for stabilization of slides in cohesive soils with a relatively high content of electrochemically active clay minerals (minerals with high water absorption and high ion-exchange capacity, as shown in Table 7-1, 7.2.2), and if these slides originate in relatively thin slip zones at a boundary between oxidizing (yellow to brown) and reducing (blue to grey) soils. Since such slides occur more frequently than is generally assumed, and since the installation of short-circuit conductors is by far the cheapest method of stabilization, it should always be taken into account within its range of applicability.

Short-circuit conductors should be placed fairly close together (usually they are arranged in a chessboard pattern, at maximum distances of 3 to 4 m; see Fig. 6-43 and 6.1.4.4.2), because due to the high electric resistance of the soil their effect does not radiate very far. Wherever highly permeable and water-bearing layers, cracks, and fissures are encountered, additional conventional drainage is necessary. Also, effective sewerage for surface waters is necessary, since otherwise—in view of the mechanism described above—the desired effect cannot be achieved.

6.1.4.4.2 Practical Applications.

The method of stabilizing landslides by reducing the electric potential differences generated at the zone of contact between two chemically different clay layers by the installation of short-circuit conductors has been introduced into practice by Veder (Veder 1957, 1963, 1964, 1966, 1968, 1972a–c, 1973; Veder and Finzi, 1962). Normally, steel rods (diameter—25 mm) are used as short-circuit conductors (ICOS, 1958).

In the following, three case histories describing landslide stabilizations with short-circuit conductors are presented. During recent years numerous stabilizations have been carried out by this method; these are collected in Veder (1972a). The soil-mechanical coefficients of a slide mass stabilized by short-circuit conductors may also be found there. It must be repeated that short-circuit conductors are not efficient where permeable, water-saturated sand or sandy-silt layers have not first been drained.

6.1.4.4.2 (I)   Landslide near St. Marein, Styria, Austria—Powerline Pole. During the spring of 1967 a slide occurred in the vicinity of a corner pole of the 200-kV powerline near St. Marein; its scarp reached the foundation of the pole and the toe of the slide was marked by wet spots. The pole itself had not yet moved.

Exploratory borings around the pole revealed, to a depth between 1.0 and 1.5 m, a layer of clay with silt embedments, with a minor flow of seepage water. Subjacent was a layer of brown, very clayey silt, followed at a depth of approximately 4.5 m by a grey-blue silt layer, also with a very high share of clay. Considering these findings, together with the position of the sliding slope, the decision was to carry out the stabilization by means of short-circuit conductors. In order to obtain exact data for the application of this method, an exploratory shaft was sunk in the area of the slide at a distance of 10 m from the pole to a depth of 8.0 m. From this shaft, which had windows cut into the concrete

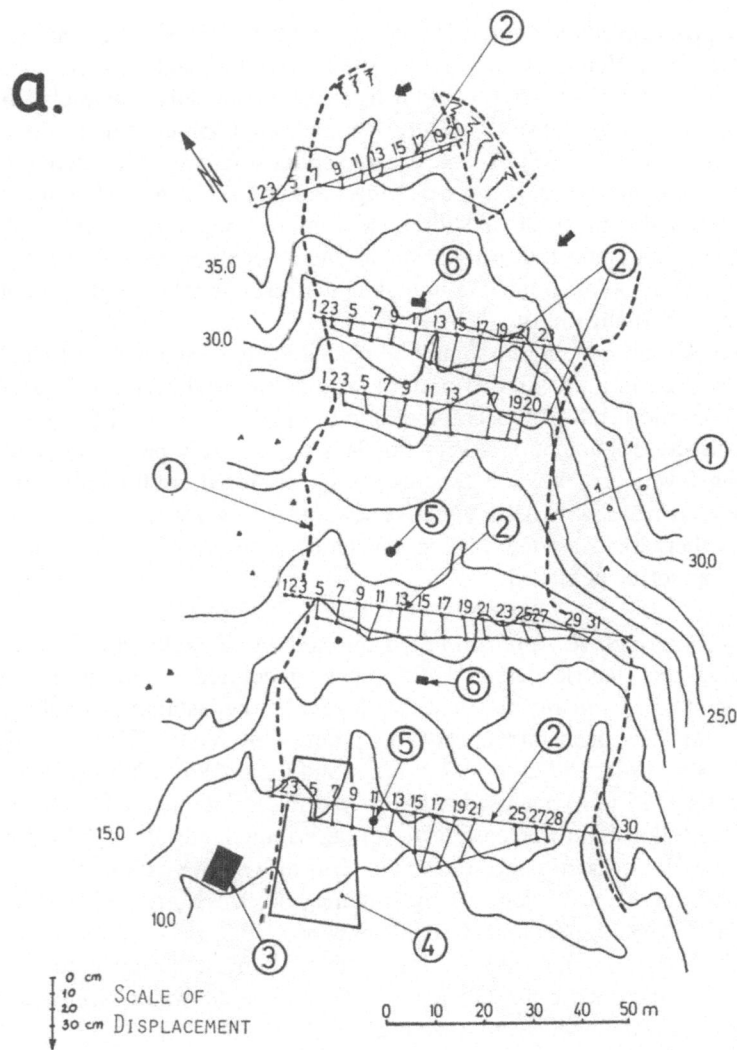

Figure 6-43 Experimental field with short-circuit conductors in a large-scale landslide in Sarukuyoji, Japan; a—site plan of landslide; (1) margin of slide; (2) profiles of observation; (3) measuring station; (4) experimental field; (5) exploratory shaft; (5) pore-water pressure gauge; b—changes in electric soil potential, pH-value and water content by the action of short-circuit conductors; c—detailed plan of experimental field with short-circuit conductors; d—measurements taken in a shaft near Oso, Japan.

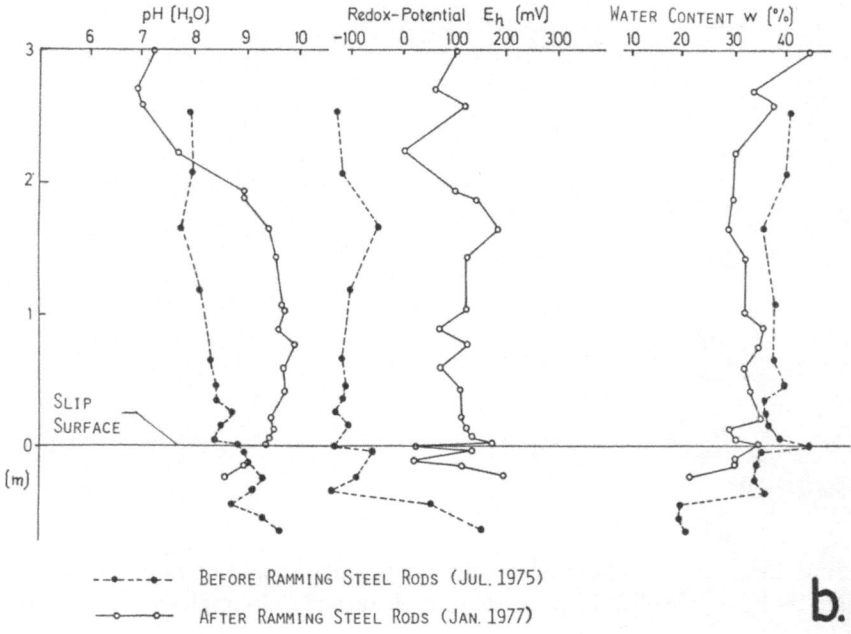

BEFORE RAMMING STEEL RODS (JUL. 1975)
AFTER RAMMING STEEL RODS (JAN. 1977)

b.

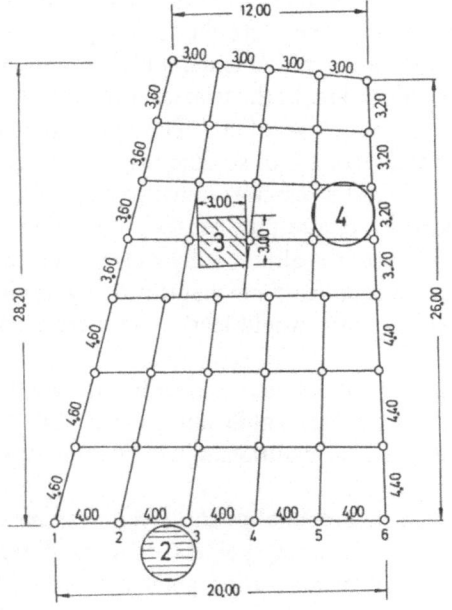

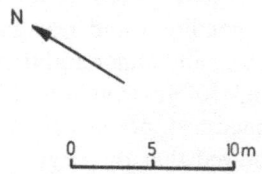

SHAFT No.	MEASUREMENT DATE
2	JULY 1975, BEFORE RAMMING STEEL RODS
3	JANUARY 1976, ONE WEEK AFTER RAMMING STEEL RODS
4	JANUARY 1977, ONE YEAR AFTER RAMMING STEEL RODS

O STEEL RODS

c.

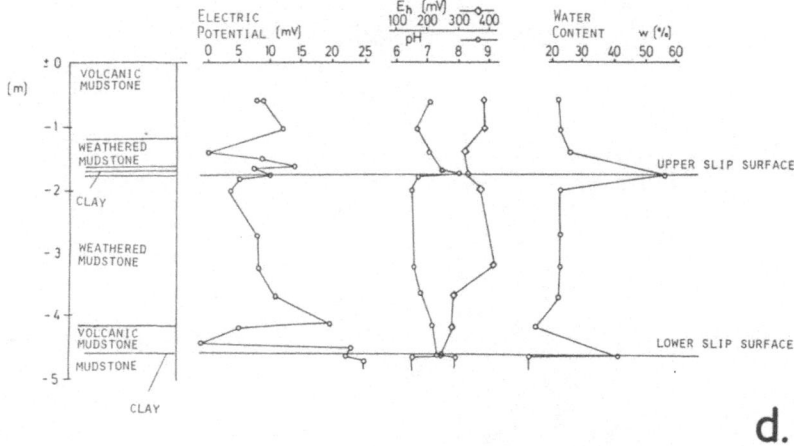

Figure 6-43

lining, the vertical gradient of the electric potential could be measured at any time. In the course of sinking the shaft, undisturbed soil samples were extracted and analyzed.

A sample taken at a depth of 4.2 m yielded the following values: natural water content $w = 54.8\%$; liquidity limit $w_L = 64.0\%$; plasticity limit $w_P = 51.0\%$; resulting consistency index $I_c = 0.71\%$. Therefore the soil was in a soft, plastic condition and was susceptible to creep movements, which would be favored by any minor additional water absorption. In the triaxial test apparatus the angle of friction was determined to be $\varphi' = 20.5°$. The consolidated, undrained test, performed at a shear velocity of 1% of sample height per minute, showed that the angle of friction of total stress was down to only 14.5°. This indicated a quickly rising pore-water pressure that causes a reduction of the angle of friction. In the slide area the soil was almost fully water saturated (saturation ratio $S_r = 97.6\%$), so that even the least mechanical or hydraulic movement, or water transport by electroosmosis, would lead to an increase of pore-water pressure.

The examination of other undisturbed soil samples basically yielded the same values, however, natural water content was significantly lower and the consistency index higher. In all samples grain distribution indicated a relatively high share of clay in the silty soil.

To determine the differences, the electric potential of the layers was measured from the shaft (Fig. 6-44). Between measuring points 3 to 6 the difference amounted to 15 mV. The voltage peak at measuring point 2 was due to the electric potential generated by the water flow in the embedments of fine sand.

Stabilization measures

(1)   Short-circuit conductors were installed in a grid, as shown in Fig. 6-45, and this stabilized the subsoil at a depth of 4.5 m by reducing potential differences and eliminating electroosmotically generated

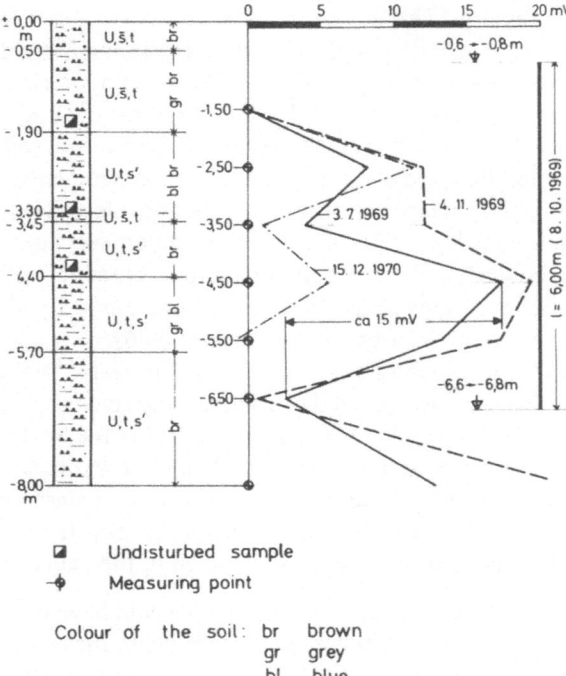

Figure 6-44 Soil profile in exploratory shaft; electric potential differences between measuring points. (For explanation of symbols used in soil profile, see 6.6.)

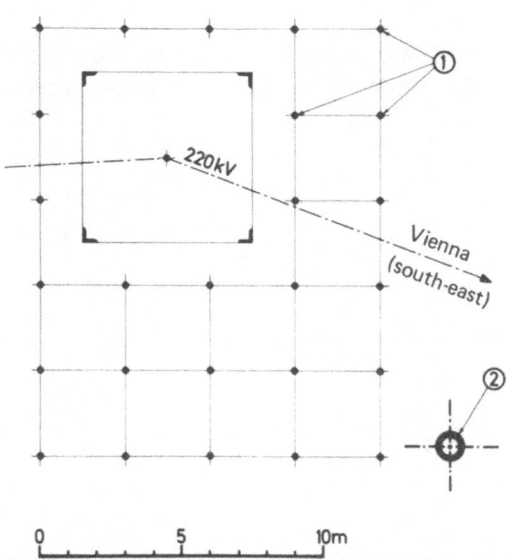

Figure 6-45 Grid of short-circuit conductors installed around pole of power line; (1) short-circuit conductors, length—6 m, diameter—20 mm, installed at distances of 3 m; (2) exploratory shaft.

water pressure. The short-circuit rods (length—6.0 m; diameter—approx. 20 mm; with couplings after every 1.5 m) were driven in so far that their tips were 60 to 80 cm below ground surface so as not to obstruct the farming of the land. Therefore, the rods reached total depths of 6.6 and 6.8 m, covering the whole zone of voltage peaks.

(2)   As the surficial layer contained sand embedments, which would quickly carry surface waters to the impermeable layers of silt and clay, a trench was dug uphill of the pole to divert seepage waters already above the grid of short-circuit conductors. Through a PVC-tube (gradient 3%) the water collected in the 1.0-to-1.5-m-deep drainage trench was diverted to a nearby ditch (See Fig. 6-46).

(3)   To prevent as much as possible the entry of water to the clay and silt layers, the slope was planted over, especially since the topsoil had been removed during planing operations and it was found that deep erosion grooves already existed underneath. Finally, it was necessary to control the efficiency of the stabilization measures against renewed movements, the gradient of the electric potential, and the water content of the soil. These measurements were taken in the exploratory shaft.

By this method, damaging excavations, which would have been necessary for the underpinning originally considered as an alternate measure, were avoided. Conventional drainage systems would have had to reach very deep to be effective and therefore would have involved much greater efforts than the method chosen. The short-circuit conductors prevented a further softening of the soil by electroosmotic transport of water, without disturbing the soil in the least. Ten years later neither movement of the slope nor soaking of the soil could be registered.

6.1.4.4.2 (II)   Landslide on the West Freeway near Viehdorf, Lower Austria (H. R. Borowicka, 1963). This slide occurred between km 191.2 and 191.8

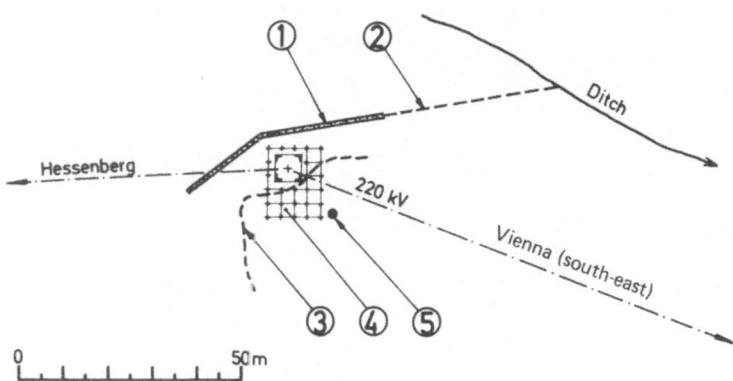

Figure 6-46   Plan of slide area with completed stabilization measures; (1) drainage trench; (2) plastic pipe; (3) head of slide; (4) field of short-circuit conductors; (5) exploratory shaft.

of the West Freeway in Lower Austria. The alignment of the roadway required a 5–7-m-deep cut into a slope with a gradient of only 2° to 3° to the northeast. The soil consisted of a superficial, 3–4-m-deep mass of yellow to grey-brown sandy-silty clay with water-bearing gravel layers, and a subjacent layer of blue to grey-black silty clay and silty clay shale.

On September 29, 1965, after cutting operations, the first movements occurred and on the following day the upper brown layer started to slide on the grey layer; at the same time staggered cracks developed in the embankment of the cut and the area above (Fig. 6-47).

To stabilize the landslide, the embankment was first graded down and drained with slope drainage; however, these did not reach down to the clay shale. But these measures did not stop the slide movements; on the contrary, they continued to spread and by summer of 1969 the whole 600-m-long embankment was in motion.

As a countermeasure, the embankment was further graded down to an inclination between 1:3 and 1:5 (vertical to horizontal) and drained. Apparently the main cause for the continued slide tendency was an electric potential gradient between the two layers (brown and grey), which was raising the water table or maintaining it at the prevailing level. Therefore, 3-, 4- and 6-m-long short-circuit rods were driven into the ground at distances of 4.5 m in the eastern area of the slide, which was moving the most. Since then, no embankment movements worth noting have been registered.

This case is a very typical example of eliminating harmful pore-water pressure at the zone of contact between clay layers of a different nature. Conventional drainage would have been virtually without effect because the inflow of water was rather minor; any drainage lines would have had to be installed very close to each other. Of course, exchanging of soil, namely, removing the upper layer and replacing it with sand and gravel, would have also been effective, but would have meant the movement of large soil masses at considerable costs.

6.1.4.4.2 (III)   Sarukuyoji Landslide, Japan (Oyagi *et al.* 1977). The Saru-kuyoji Slide took place in the western part of the Niigata District, one of the

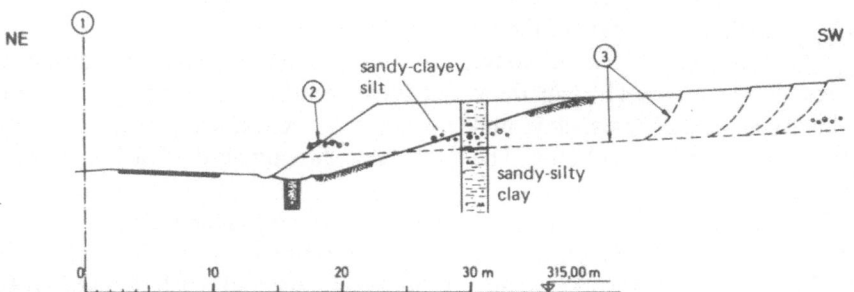

Figure 6-47 Section of slide area; (1) axis of motorway; (2) water outlet; (3) slip surfaces.

areas in central Japan with the heaviest snowfall. The landslide is about 1,700 m long and has a mean width of 250 m. The Institute of Public Works, headed by Watari, had chosen this landslide as an experimental field for development and improvement of methods to monitor and stabilize slides (see 6.3, Table 6-2). Among other measures, a grid of 44 short-circuit conductors was installed and the effect observed from an exploratory shaft (6.1.4.4.1, Fig. 6-43).

History of the slide. Apparently, landslides had always occurred in the area and a legend tells of a slide in the year 1100 that destroyed the farmlands, whereupon a Buddhist priest was buried there as a human sacrifice to avert the danger of landslides. In 1937 a casket containing his bones was found. In 1900 a major slide occurred at the head of the slope and a flow of earth masses covered almost the whole expanse of the slide area. The middle section of the slide became active again during the great Kanto earthquake (7.9 on the Mercali scale), one of the most catastrophic earthquakes in the history of Japan; the epicenter was 240 km south of Sarukuyoji (1923). Starting in 1943, a creep movement was observed near the foot of the slope; some houses and a large area of farmland were damaged.

Since 1970 the upper part of the slide has also been in motion. Currently, the slide movement in the middle section covers about 3 m per year.

Geology and topography. The 571.6-m-high, flat Jogayama Mountain is surrounded by a number of slides. These take place in the marl of the Miocene Teradomari formation, the most slide-prone soil in Japan.

The marl strikes with a 20° gradient to the northeast and dips at a 30° gradient to the southeast. The slide extends from a height of 450 to 500 m towards the southwest, down to the Okuma River flowing at the foot of the mountain.

Structural characteristics. The slide represents a creep movement of unconsolidated soil masses. This material may stem from the head of the slide area, from where it has come down during previous slides.

The slide movement takes place at approximately a right angle to the strike of the layers; its thickness varies between 12 and 15 m at the head, 5 m in the middle, and 10 to 15 m at the foot.

Near the experimental field the surface is slightly undulated, with a mean gradient of 10°, and two steeper sections with gradients of 25° and 35°. Numerous cracks are evident in the slide area. The sharp edge of the scarp at the head is probably part of the failure of 1900.

Soil properties. The main body of the slide consists of two layers, an upper layer of brown clay (1–3 m thick and impermeable), followed by a 2–3-m-thick layer of dark-grey clay, under which a 2–3-cm-thick slip surface with 50% water content was found. The latter rests directly on the hard, firm, non-sliding marl.

The mineralogical structure of the upper, brown clay layer and the subjacent grey one is virtually identical; both soils have a predominant share of montmorillonite. The 50% water content of the slip surface is about 10% higher than that of the grey clay.

Thorough measurements of water content, redox potential and pH-value were carried out before and after the installation of short-circuit conductors (see 6.1.4.4.1, Fig. 6-43).

In the upper, brown layer Leptothrix and Sedrothrix bacteria were found to a depth of 3 m. Investigations are under way to established whether these bacteria take part in the conversion process from blue to brown clay.[1]

Recent movement measurements. Numerous measurements were taken in the slide area, using extensometers and inclinometers, and in addition movements were monitored by measuring points installed at the surface.

The longitudinal deformation amounts to 1 to 3 cm per year, being higher in the steeper than in the flatter parts of the slope. The slide movements were influenced by heavy rainfall in August, the Bain Typhoon, and by melting in spring after heavy snowfalls (up to 4 m) in winter. The ground-water level plays a major part in slide movements. The water from precipitations, entering through numerous fissures, penetrates only the upper brown layer but does not reach the slip surface at the depth of about 5 m.

Whenever the ground-water table rises to only 0.7 m below ground surface, the slide movement accelerates; it quickly slows down again as the water table sinks and stops altogether when it drops to 1.5 m below the surface.

The clay of the slip surface has the following strength values in relation to water content:

Natural Water Content	Shear Strength
50%	4 kN/m²
45%	6 kN/m²
40%	10 kN/m²

This means that a small decrease in water content considerably increases shear strength.

Stabilization measures. Surface as well as subsurface drainage was carried out, especially in the brown clay. Furthermore, a bore-pile wall was constructed in the eastern part of the slide. Both measures proved effective.

By the installation of short-circuit conductors water content in the experimental field was reduced from 50 to 40%, and this led to a marked slowing of slide movements. During the period from December 1976 to June 1977, the movement in the experimental field amounted to 20 cm, as compared to 60 cm outside this field.[2] Observations are continuing.

---

[1]Shortly before going into print, the author received a letter from Dr. F. Blümel, Bundesanstalt für Kulturtechnik und Bodenwasserhaushalt (Federal Institute for Cultivation Techniques and Ground Water Balance), Petzenkirchen, Austria, in which he reports on his research and that of Professor H. W. Ford of the Institute of Food and Agricultural Sciences, Florida University, U.S.A. Leptothrix are in fact iron bacteria which have an oxidizing effect upon bivalent iron compounds, for example, those in grey or blue clay. By the action of Leptothrix bacteria, more oxygen is present in brown clay, since it contains by far more trivalent than bivalent iron compounds.

[2]The short-circuit conductors (steel rods, diameter—25 mm) were driven by the Taisei Company to depth of 7 m. Here Messrs. Aoto, Jihoshi, Kaneko, Nakao, and Suzuki have contributed special personal efforts. The same team has successful stabilized a 60-×-35-m field near Oso, in the vicinity of Kobe, by the same method.

6.1.5   Increase of Internal Friction—Solidification of Soil

6.1.5.1   Solidification by Grouting (Claquage, Soil Fracturing)

6.1.5.1.1   Overview. Grouting serves to solidify and seal the soil because the grout enters the soil pores. Cement suspensions are used for sand and gravel soils, chemicals for soils containing fine sand. The use of sleeve tubes is recommended because they make it possible to direct the grout to the exact spots where it is required. Grouting has been used with success in the solidification of loose soils, the stabilization of landslides, but also when advancing tunnels, e.g., during the construction of the subways in Vienna, Milan, and several cities in the German Federal Republic.

With very finely grained soils, such as loess, silt, silty clay, and clay, the grout does not penetrate into the pores. In these cases first vertical and horizontal fissures are opened up by temporary thrusts of 5 to 30 bar; then these are filled under a lower pressure with the solidification material, e.g., cement mortar (Fig. 6-48). By this method, called "claquage" or "soil fracturing," a net of reinforced branches and ribs is created; if the distance and sequence of the points of application were properly chosen, this may considerably increase stability conditions

6.1.5.1.2   Practical Applications

6.1.5.1.2 (I)   Hart, near Gleisdorf, Styria, Austria. Near Hart, 8.5 km north of Gleisdorf, a relatively flat slope of clayey-sandy silt from the Pannon

Figure 6-48   Control of grouting (claquage) by excavation.

was creeping towards the creek. On the road connecting Hart and Gleisdorf, longitudinal cracks were developing again and again, and had to be filled. This case was similar to the one described in 3.5.1.4, where streaming water triggered a creep movement although water supply was relatively minor.

Drainage, as described in 6.1.4.1 and 6.1.4.2, would have been a possible alternative, but the soil was relatively impermeable and therefore the drainage lines would have to be arranged very close to each other, rendering this method uneconomical. Drainage by electroosmosis would have been quite expensive too. Short-circuit conductors, as described in 6.1.4.4, were also out of the question, since this was not a phenomenon at the zone of contact.

Consequently it was decided to grout the approximately 530-m-long and 15-to-25-m-wide area by ramming in grouting tubes with lost tips at distances of about 6.0 m to depths between 4.5 and 12.0 m (Austrobohr, 1978).

Into these tubes was pressed a mixture of cement and ground limestone, to which stabilizing and plasticizing agents had been added. The ratio of water to solids was kept so low that the mixture was just able to flow but would contain only very little free water after setting. Grouting operations created a pore-water overpressure which disappeared again after 10 to 30 days; during that period some minor soil movements still occurred. The set cement mortar had a strength of $\sigma_B = 500$ to 900 N/cm^2.

From June to November 1969, an area of 8,500 m^2 was treated with 7,000 m of grouting tubes—which were withdrawn again—using up 2,000 tons of solid material. Shortly after stabilization work was completed, creep movements stopped and since then the slope has been fully at rest. Probably the chemical reaction between lime and silt, as described in 7.2.2, also played a part in the solidification process.

6.1.5.1.2 (II)   Mürzzuschlag Tunnel, Styria, Austria. In the digging of tunnels or intersecting of soils that tend to subsequent cave-in, the claquage method has the disadvantage that there is no true guarantee regarding the volume of grout required and its success; therefore, the applicability of this method must be investigated with large-scale tests. Grouting may be carried out from the excavation site, but this has the disadvantage that grouting work slows down the progress of the tunnel. If the tunnel lies at a shallow depth, this can be avoided by carrying out the grouting from the ground surface. When the Ganzstein Tunnel for a 7.5-m-wide roadway was dug near Mürzzuschlag, the greatly feared quartzite–dolomite rockflour from the Semmering-Mesociocene,[3] mixed with water, flowed into the tunnel and greatly impeded excavation operations. An about 10 m wide surface subsidence developed. As a countermeasure, the sidewalls were protected with bore-pile walls and the groundwater table was lowered on the uphill side; the latter was carried out ahead of

[3]Quartzite-dolomite rockflour characteristics:

	d_{10}	d_{60}	w	w_L	w_p	φ'	c'
Without clay:	0.005 mm	0.02mm	17%	—	—	46°	5 kN/m^2
With clay:	in clay range	0.008 mm	18.7%	22%	15%	27°	5 kN/m^2

the excavation work in the tunnel. The tunnel roof was secured by pressing in grout, which set properly; as the tunnel progressed it was found in the area of the roof as a 1-to-3-cm-thick concrete slab. Finally, the excavation work was carried out in sections—first the sidewalls and then the roof. These measures made it possible to dig the tunnel without further problems.

6.1.5.2   Drainage and Consolidation by Electroosmosis (after Casagrande)

6.1.5.2.1   Overview
by F. Hilbert
The theoretical basis of electroosmotic drainage is presented in 7.2.2.4.2. and a detailed description of its practical application in the stabilization of slide-prone slopes may be found in 6.1.5.2.2. Here only a few general aspects are discussed.

In principle, electroosmotic drainage may be used with two different objectives: On the one hand it may aim at only temporary lowering of the groundwater table, or a temporary prevention of slides (generally to keep construction pits open, or while digging a tunnel, or during cutting operations, until the construction is completed); on the other hand, a permanent stabilization of slopes may be intended. Depending upon the desired results, the method has to be applied differently and in particular it must be taken into account that not all types of soil are equally well suited for this method.

To keep a construction site open only over a certain period of time, electroosmosis will work well in clayey silt and sandy-clayey silt soils with a low plasticity coefficient, relatively small content of electrochemically active clay minerals (not more than 15 to 20%), and a grain distribution that corresponds approximately to Fig. 6-49. Electrochemically active clay minerals in this sense are those shown in Table 7-1 (7.2.2) as having high swelling and ion-exchange capability.

To obtain permanent consolidation by electroosmotic drainage, a sufficient permanent consolidation pressure and/or an electrochemical change of soil properties is necessary (see 7.3.3.2 and the following descriptions of soil changes by electroosmosis). Accordingly, the higher the content of electrochemically active clay in the soil, the easier is permanent consolidation obtained. For consolidation processes based on strictly chemical effects, even if the chemicals are introduced into the capillary system by electroosmosis, other considerations apply (see 6.1.5.3.1).

If possible, electroosmotic drainage should begin before excavation work is started, because simultaneous soil pressure considerably improves the effects of the drainage and thereby speeds up the consolidation.

The homogeneity of the soil is also of special significance in an application of electroosmosis. If soil with a low permeability coefficient (silt, loam, clay) is stratified, or if the soil has a high permeability coefficient, or if there are water-bearing cracks and fissures, difficulties will be encountered. The clayey-loamy portions of the soil may again absorb water from permeable layers, so that no

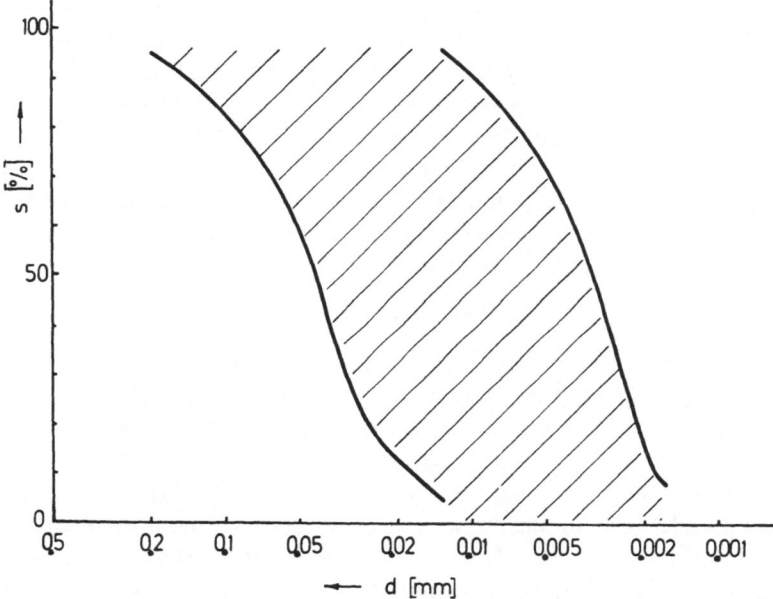

Figure 6-49 The shaded section defines the range of sieve-analysis curves of sandy-clayey silt and clay soils, where a temporary lowering of the ground-water table or drainage by electroosmosis may be economical (after Casagrande); s—percentage passing sieve; d—grain diameter in millimeters.

real drainage but only a flow-through of water results. Consequently, field and laboratory tests always have to be carried out before electroosmosis is applied, except where it is used under time pressure as a last recourse.

As a field test it is usually sufficient to sink a filter well with a cathode (well point) and to drive in two or three anodes at distances of 1 to 5 m (e.g., steel rods or water pipes). The well and the anodes should reach somewhat below the soil layer to be drained. If the inflow of the well, as determined by pumping, can be increased markedly by applying d.c. tension with moderate electrical power between the cathode and one of the anodes, and if a good drainage effect and consolidation is achieved around the anode, then it may be assumed that electroosmosis will be effective.

Of course, approximate soil characteristics have to be determined in laboratory tests from undisturbed soil samples, before the field test is set up. The hydraulic coefficient k_h (electroosmotic drainage can only be applied if $k_h < 10^{-7}$ to 10^{-8} m · s^{-1}), the electroosmotic coefficient k_e, the specific soil resistance ρ, and the plasticity coefficient decide if electroosmotic drainage is applicable. Specific resistance may also be determined by a field experiment for two electrodes from

$$\frac{U}{I} = \frac{\rho}{\pi \cdot L} \ln\left(\frac{d}{r}\right)$$

(U = tension; I = voltage; L = length of electrodes below ground-water table; d = distance between electrodes; r = diameter of electrode). The quantity of extracted water may be approximated from $Q = k_e \cdot \rho \cdot I$.

The plasticity coefficient should be so low that a slight decrease of water content will result in a sufficient consolidation, because, on the one hand, electroosmosis will drain off only a limited part of the water (almost up to the point where plasticity vanishes), and on the other hand, the closer one gets to the limit of drainable water the slower and more expensive the process, because collection decreases exponentially with time:

$$Q = W_0[1 - e^{(-k_e \cdot E \cdot t)}]$$

(Q = quantity collected; W_0 = pore volume; E = field intensity).

In practical drainage, d.c. potentials in the range of 15 to 150 V are used for distances of 1 to 5 m between anode and cathode, with maximum field intensities from 10 to 100 V $\cdot$ m^{-1}. The cathodes are usually arranged in one or several rows, at a right angle to the inclination of the slope; the anodes are placed parallel, also in one or two rows, and between the rows of cathodes if several of them are set up. Less often, a chessboard pattern of cathode wells and anodes is used. The applied voltage should be kept so low that the desired reduction of water content is just achieved within the available time; higher voltages use up more power and also increase the danger of drying out the soil around the anodes. This drying-out process may be caused by the heat generated by the electric current, but it may also happen that water is extracted more quickly than it rises up through capillary action. Dryness always leads to a large increase of soil resistance, which may stop the electroosmotical process. Consequently, a voltage or potential must always be chosen that causes, at the most, a 5 °C temperature increase at a distance of 10 cm from the anode.

It is always necessary to pump or drain off the water from the cathode wells, if the plan of zero pressure is to be lowered effectively. The combination with vaccum-drainage, as used in Case History 6.1.5.2.2, was found to be especially effective.

If electroosmosis is used to obtain permanent soil consolidation, considerably higher electric power has to be applied (up to 30 A/m^2 of the area covered by the row of electrodes), whereby an electrochemical transformation of the clay minerals sets in after longer periods of current flow (decreasing pH-value around the anodes and migration of higher-valency ions into the soil, formed by dissolution of the anode materials—iron or aluminum).

6.1.5.2.2 Practical Application—Kootenay Channel, British Columbia, Canada.

Around 1930, Casagrande introduced the basic idea of electroosmotic drainage and consolidation into soil-engineering practice. A more recent application (Griffin, 1972) is described in the following.

The 30-m-high slope of very loose, sensitive silt uphill from the Kootenay Channel power plant in British Columbia was consolidated with five rows of electrodes running parallel to the slope over lengths of 100 to 300 m. The electrodes reached down to the bedrock (up to 30 m in depth) and were installed in pairs at distances of 10 m. The cathodes were wellpoints and the anodes consisted of 56-mm-diameter steel tubes jetted in at a downhill distance of 3 m.

The wellpoints were contructed as follows: First, steel tubes (diameter—25 to 30 cm) were jetted in to depths of up to 30 m. Then the soil was removed from these tubes, sand was filled in and the tubes were pulled out, so that a column of sand remained in the soil intersecting the ground-water table. This sand column contained a pipe (diameter—50 mm), with an inner piper (diameter— 32 mm). Water was forced under high pressure through the outer pipe and conducted through a Venturi tube into the inner pipe, where it rose again and also forced up the ground water. The water was drawn to the wellpoints by a 100-to-150-V current, and 160 liters/min were pumped from the 30 wellpoints of the uppermost row.

After about a week, success became evident: The slope, that was until then only stable at a 1:5 inclination (vertical to horizontal) could be steepened to 1:2, so that the next row of electrodes could be installed. Six months later the whole slope was stabilized. During this period a 100-kW d.c. motor and two 6-in-centrifugal pumps were in operation.

In order to test slope stability after the electroosmotic treatment, two exploratory trenches were dug close to the foot of the slope, reaching to a depth of approximately 10 m. The sidewalls of these unbraced trenches remained absolutely vertical over a long period. Therefore it could be safely assumed that the consolidation by electroosmosis had a definite effect.

The soil-mechanical coefficients had undergone the following changes:

(1) By agglomeration the silt component had decreased from 55–25% to 18–0%.

(2) At a water content of 30%, the undrained, consolidated shear strength had increased from 0.14–0.56 MN/m^2 to 0.4–1.00 MN/m^2.

(3) The gradient of Mohr's envelope curve, originally 27° to 32°, increased within the 6 months of treatment by electroosmotic drainage to 35°.

It may be safely stated that in the presence of water-saturated silty soils (share–20 to 55%), drainage by electroosmosis today seems the only possible method to stabilize slopes susceptible to slides.

6.1.5.3 Consolidation by Compounds of Magnesium, Calcium, Aluminum or Iron

6.1.5.3.1 Overview
by F. Hilbert
In principle, replacement of monovalent cations in loam and clay soils by multivalent cations causes consolidation (7.2.2.3). In practice this ion exchange may be carried out in two ways: either by mechanical methods or by electroosmotic injection of ions. Chemical consolidation is also possible.

(A) *Mechanical adding of solid or dissolved calcium and magnesium compounds.* So far only slide stabilization by mechanical mixing-in of unslaked

caustic lime, as described in 6.1.5.3.2., has been tested and is approved in practice. In addition to consolidation by ion exchange, the lime dries up the soil,
causes purely mechanical consolidation by formation of solid calcium compounds (e.g., calcium carbonate by reaction with carbonate contained in rain
water), and increases the electrical conductivity; this increased electrical conductivity of the pore water increases the hydraulic permeability of the soil and
consequently improves drainage. Due to these chemical consolidation effects,
this method is not only suited for loam and clay soils but also for silty and
sandy soils with low hydraulic permeability; these require a somewhat higher
share of lime. Other methods, such as mixing the soil with ground limestone,
magnesite or dolomite, have so far only been discussed in the literature.

The treatment with saline solutions to stabilize slides has until now not been
taken into consideration, although the application of calcium sulfite lye from
paper mills has proved quite effective in stabilizing macadam roads.

(B) *Electrochemical injection of aluminum, iron, calcium, and magnesium
ions*. During drainage and stabilization after L. Casagrande (6.1.5.2), the commonly used iron or aluminum anodes are dissolved and the stabilization in the
vicinity of the anode takes place at least in part by migration of higher-valency
iron or aluminum ions; the same applies to the short-circuit conductor method,
after Veder (6.1.4.4). So far no other practical applications are known,
although it is, for instance, under discussion to mix calcium chloride into the
anode bedding, which should cause electrolytic migration of calcium ions
towards the cathode.

(C) *Chemical consolidation*. In principle, a purely chemical consolidation
may be achieved by treating the soil with solutions of chemicals that form solid
compounds after injection, either by reaction with components of the soil or by
transformation (e.g., flocculation of gels); the solids are thus formed just in the
soil pores and cause agglutination of the soil particles. Here again either purely
mechanical injection or electrolytic infiltration is possible. In the latter case the
solution is usually put under pressure too, but this pressure will be much lower
than that required for purely mechanical injection. As anodes, perforated steel
tubes are installed in a permeable packing and filled with the solution to be
injected. A solution proposed for injection is 40% $Mg\,[SiF_6]$ (Fritsch, 1976).

Practical application of chemical consolidation is rather limited by high
chemical and labor costs. The method is thought to be economical only under
extraordinary conditions.

6.1.5.3.2 Practical Applications

6.1.5.3.2 (I)   Mooskirchen, Styria, Austria (Homann, 1975). A recent
example of the stabilizing of a cutting slope with lime is described here. Where
the South Freeway ascends towards the Pack, immediately after the road from
Mooskirchen connects with it, a cut was made up to 10 m deep in tertiary silt,
which contained, in part, sand and marl (Fig. 6-50). A 2:3 inclination (vertical
to horizontal) was chosen for the cutting slope and over the first 30 m it

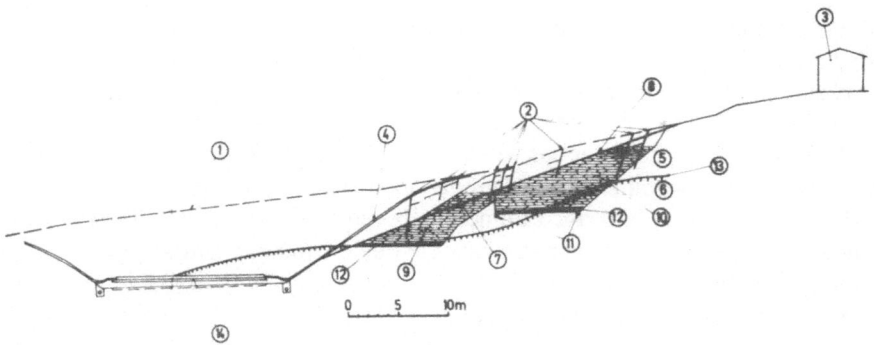

Figure 6-50 Section of slide area with stabilization measures; (1) original ground; (2) scarps; (3) building; (4) designed slope; (5) weathered sandy-clayey silt (loam); (6) marly, silt-mudstone; (7) first grading-down of slope; (8) completed slope; (9) first stabilization by lime; (10) second stabilization by lime; (11) drainage; (12) filter gravel; (13) potential slip surface; (14) motorway.

remained stable. In the further course of grading the slope, slickensides with a 60° gradient to the roadway, reaching to the potential slip surface, were encountered exactly in the zone downhill of a residential building. A further grading down of the cutting slope, as ordered at first, did not stop the movement already under way and the head of the slope, which had already been set back by cutting operations, moved critically close to the residential building. Therefore a stabilization had to be achieved quickly and it was decided to consolidate the soil with lime and then to rebuild the slope, from the toe upwards, by using the material available on the site. This method offered the advantage that it could be carried out quickly, in only 2 days, and involved the movement of only 1,700 m³ of soil, the use of a bulldozer, a part-time loader, two trucks, and a lime stabilization unit consisting of a ground cutter and a spreader.

The soil material was obtained by bulldozer, which at the same time undercut the slope and the slickensides. The reconstruction of the new, flatter slope was carried out in two stages (Fig. 6-50, Stabilization I and II). For the lime stabilization first a level course, wide enough to accommodate the ground cutter and the spreader, was graded. Then soil was taken from the purposely.oversteepened slope, mixed with lime and spread in layers on the level course, thus gradually building up this terrace. At the middle of the cutting slope a water-bearing layer was encountered, which was evidently responsible for the slide. This layer was drained on the slope face with gravel and the water was temporarily diverted down the slope. The discharge amounted to approximately 1 liter/min. By this prompt action, which involved only a minor investment of labor, the slide was prevented from reaching the house. In Styria, slides are usually stabilized by stone work; this method would have involved 1 week of work and the complete removal of the slope material. The material would have had to be disposed of as useless and replaced by stones. However, the method described here could only be carried out because it was possible to flatten the slope to a 20° gradient. Another precondition was that the material on the site was suitable for stabilization with lime.

6.1.5.3.2 (II) Construction of Lime Piles. Brandl (1973, 1976a) first reported on the stabilization of sliding slopes with lime piles. Vertical holes (diameter—8 to 50 cm) are rammed or bored to depths of about 10 m and filled with caustic lime, which is compacted by continuous tamping or by compressed air. In view of the relatively low diffusion velocity of calcium ions in the mostly cohesive soils, small diameters and short distances within the grid are preferable.

Broms and Boman (1976) describe a different type of lime pile (actually lime–loam piles). In this method, clay ($w \cong 100\%$, $w_L \cong 90\%$, $c_u = 12$ kN/m²) is stabilized by mixing in unslaked lime, together with other suitable additives, without removing the soil material. By means of a quirl, mounted on a mobile boring rig, a lime pile (diameter— 0.5 m) can be drilled within about 15 min to a depth of up to 10 m, whereby the stabilizing agent is mixed in at a ratio of 5 to 8% of the dry unit weight of the soil. Two months after its completion, a test pile in clayey silt had achieved half of its final strength; after about 1 year it had fully set and had the 50-fold strength of unstabilized clayey silt or the 5-fold strength of postglacial clay. It is recommended that thorough laboratory tests be carried out before a large-scale application of this method, since the effectiveness of lime piles varies with subsoil conditions.

The authors report on tests carried out in a 5.2-m-deep ditch, where overlapping lime–loam piles were used in place of sheet-pile walls. This method has so far not been used for slide stabilization, but in my opinion it might be very effective.

This type of lime piles offers the advantage that the slide mass may be stabilized to depths of 5 to 10 m; that is deeper than with the method described in the preceding Case History (6.1.5.3.2 (I)).

6.1.5.3.2 (III) Sonnenberg Road, Switzerland. For the sake of an optimal design of the alignment, the route of National Road N2 had to cross a slide-prone slope uphill of the village Itingen, Canton Basel, Switzerland, called the *Sonnenberghang* (Jedelhauser, 1970).

The 300-m-long embankment required a 2–9-m high fill above the original ground surface, with a 30-m-wide crown. The *Sonnenberghang* was known to be a sliding slope. Subsoil conditions were the following: On top of a layer of opalinus-clay, with a dip parallel to the slope surface at a 1:4 inclination (vertical to horizontal), rested a layer of residual loam (sandy-clayey silt) of varying thickness.

From the uphill impermeable Sowerbyi and Murchsonae layers, the entering precipitations would flow to the surface of the opaline-clay and there run down, softening the zone of contact between opaline-clay and residual loam.

Under these circumstances, placing the fill on top of the residual loam would have endangered the stability of the embankment and it was necessary to excavate the loam completely (mean depth—6 m), down to the surface of opaline-clay.

This decision was reached in view of the fact that the loam had the following relatively low coefficients:

$$w = 18\%, \quad \varphi' \text{ present } = 20°-22°$$
$$w = 22\%, \quad \varphi' \text{ present } = 16°-18° \quad \gamma \text{ present } \cong 1.95 \text{ t/m}^3,$$

while the safety calculations for the embankment with slide circles yielded the following required values ($\eta = 1.5$): in the upper part of the embankment φ' required $= 22°$, γ required $= 1.9$ t/m^3; in the lower part, φ' required $= 32°$, γ required $= 2.0$ t/m^3. The values for the lower part were not quite achieved. Therefore this material, if not consolidated by lime, was not suitable as fill for an embankment with a supporting function.

The residual loam was then stabilized in sections with caustic lime and used as fill for the embankment core on top of the opaline-clay. The dose of unslaked lime mixed into the soil (Fig. 6-50) was 35 kg/m^3 for the approximately 6-m-high embankment core, and 18 kg/m^3 for the approximately 5-m-high upper part. This dose was designed for an 18% water content of the loam; where it was higher, the dose had to be increased proportionately (Fig. 6-51).

Since the construction pit could be opened only by 10 m at a time, so as not to trigger slides, the excavation had to be carried out not parallel but at a 45° angle transverse to the slope. The residual loam was mixed by layers of 25–27-cm thickness with lime and then the stabilized material was used to fill the embankment core. Prior to filling the embankment, it was necessary to cover

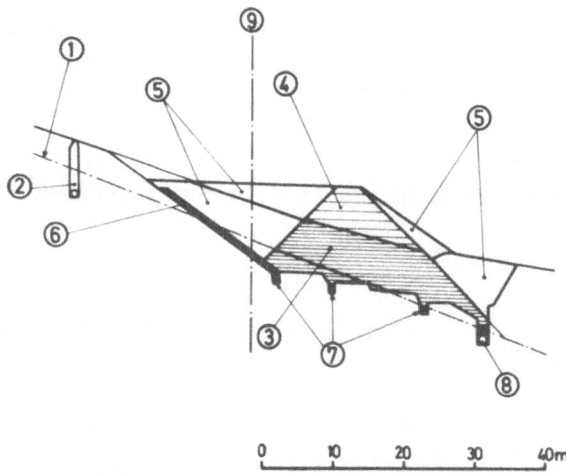

0   10   20   30   40m

Figure 6-51 Section of slide area with stabilization measures; (1) slip surface; (2) slope drainage carried out before start of stabilization work; (3) soil stabilized with lime (35 kg/m^3), $\varphi' = 30°$, $\gamma = 1.9$ t/m^3, 95% proctor density; (4) soil stabilized with lime (18 kg/m^3 lime), $\varphi' = 22°$, $\gamma = 1.9$ t/m^3, 95% proctor density; (5) unstabilized soil, compacted by layers; (6) gravel layer (30 cm); (7) seepage trenches inclined to main drainage; (8) main drainage; (9) axis of motor route.

the uphill slope with a 30-cm-thick drainage layer and to construct seepage trenches leading to a main drainage, running parallel to the base course of the embankment. The 30-cm-high layers of fill were compacted by dumping rolls or vibrating plain rolls.

Before starting out with the next level of excavation, the natural water content of the soil was analyzed in the field laboratory. Then the contractor was advised regarding the quantity of lime to be spread per square meter.

In the city laboratory the achieved shear strength was tested with samples from the construction site. These tests yielded the following mean values: Dose of CaO—35 kg/m^3; density $\gamma[t/m^3]$—2.00; angle of friction φ'—30°. To complete the embankment body, lime-stabilized loam was spread on both flanks and compacted.

Before starting the excavation and filling work, a drainage was installed parallel to the slope, uphill of the construction site. It reached by at least 2.0 m below the surface of opaline-clay. The necessity for this drainage was clearly demonstrated because in the section where the projected drainage was omitted, a slide occurred during construction work.

Finally it should be pointed out, that the following alternatives were under consideration before selecting the stabilized loam as embankment fill: (a) using the opaline-clay or Murchison-rock available on the site or (b) using the Blagdine or Hauptrogenstein material excavated during the construction of a nearby tunnel. However, it was determined by test fillings that these materials lost their rocky characteristics within a short time and thereby the high angle of friction required of the fill for a supporting embankment body. In addition, these materials did not achieve an acceptable bearing capacity in spite of thorough compacting.

In short it may be said that the method of stabilizing the residual loam by lime in order to increase its angle of friction, as well as the process chosen for carrying out filling work, proved fully efficient.

6.2 The Morphology of the Terrain is in Part Significantly Changed

6.2.1 Improvement of Stability Conditions in Sliding Slopes or Embankments

6.2.1.1 Down-Grading to Relieve Slope Head of Load Pressure—Headrace, Rosegg Power Plant, Carinthia, Austria

The experiences with the channel for the headrace of the Rosegg Power Plant, which is 1.7 km long and cuts with a mean depth of 28 m (maximum depth 50 m) into the terrain (see Ludwig, 1975) were similar to the phenomena described in 3.2.1.

Below a layer of moraine deposits of varying thickness, this channel cut into mica slate bedrock which caused the greatest stability difficulties since it partly reconverted into the clay from which it once had formed.

In order not to start out with a gradient that was too flat, which would require a lot of expensive planing operations, first, a 2:3 inclination (vertical to horizontal) was chosen. A 5-m-wide berm was excavated 11 m above the bottom of the channel, which was 8 m wide. This means that construction work was carried out after Terzaghi's observation method, purposely not aiming from the start at the safest and most expensive solution, but adapting, in the course of construction work, definite measures to local requirements. This strategy is especially recommended with bedrock containing layers capable of swelling and with soil-mechanical coefficients that cover, after Borowicka, a very wide range ($\varphi' = 18-25°$, $c' = 0-5 \text{ kN/m}^2$).

During excavation operations, started in the summer of 1972, the slope, which could only be broken by grabs after light blasting and was supported by apparently solid rock blocks, was considered stable; there were, in addition, only a few moist spots.

This stability proved to be deceiving. About 2 to 3 weeks after the cutting slopes had been compacted, scarps developed on the left bank uphill from the berm; the right bank remained stable, since there the layers dipped towards the slope.

First, tension cracks developed, progressing from bottom to top, and then slope-parallel failures occurred at depths of about 3 m, revealing slickensidelike shiny slip surfaces. Apparently a weakening of the diagenetic cohesion (see 3.3.2) played a major part.

Based on past experience, the slope was graded down to 1:2 and provided with a 4.5-m-wide berm at about the same level as in the original design.

In addition, bore-hole drainage was installed uphill of the berm to reduce pore-water pressure, and grass was sown for further drainage and stabilization.

In the spring of 1978 isolated slide movements started again in the upper third of the channel cut, partly just above the berm, at the borderline between the relatively permeable moraine and the subjacent, relatively impermeable mica slate.

There was a higher water-inflow to these zones, but the quantities as such were so minor that even precisely placed cased drainage borings were not effective; the water only dripped from the pipes sunk at right angles to the slope surface. Consequently, three additional stabilization measures were carried out.

(1) Bundles of willow rods were placed in surficial trenches at the zone of contact between moraine and mica slate. These withdrew water from a wide area and the roots also stabilized the soil.

(2) "Krainer Walls" (walls formed by wooden or concrete beams placed one upon the other with regular spaces, vertically and parallel to the slope—6.2.9.5 (I)) were constructed where major scarps had developed at the zone of contact between moraine and mica slate. These walls retained the soil and still allowed minor movements; they also did not constrict drainage.

(3) Where even these measures were not sufficient, the sliding mass was retained over a length of about 400 m by anchored concrete slabs.

These concrete slabs of the "Piz Badura Zyklops" type (Badura, 1976) are 1.20 m high and either 1.20 or 2.40 m wide, with a mean thickness of 0.20 m and a 0.7-m-wide foot. They have the shape of cantilever retaining walls, and either rest with the 0.7-m-wide foot on a concrete beam sunk into the soil or are bolted in two to three rows, one on top of the other, to the slope.

The slabs, which have a great number of weep holes, are pressed at an angle of 30° to vertical against the cutting slope covered with a filter layer by using concrete flanges (height—0.50 m; width—0.80 m). In mica slate the pressing is carried out by rock anchors of the Dywidag Thread Anchors type (tension style, type St 835/1030(85/105), diameter—32 mm) at an angle of 20° below horizontal.

The 15-m-long frictionless anchors, with a 5-m-long adhesion zone (diameter—32 mm), are designed to take up 500 kN and may be retensed if required. The distances between anchors varied between 3.20 and 1.60 m, depending upon the earth pressures of 100, 200, or 300 kN/m.

The slope uphill of the anchored concrete wall, which now has a 1:3 inclination, has not moved since the above stabilization measures were carried out. For further stabilization the slope was planted with bushes.

6.2.1.2 Replacement of Too Heavy an Embankment Fill (e.g., next to a bridge)—Krummbach Bridge, Styria, Austria

In January of 1967, shortly after the completion of the up to 7 m high embankment leading to the bridge abutment facing towards Vienna, a slide occurred in this area which badly damaged the abutment. The slide was probably triggered by pore-water overpressure resulting from the high water saturation of the surficial layer from precipitation, and the load of the embankment fill. The main movement of the slide was directed towards the axis of the bridge at a gradient of 45° to 60°. It could be assumed that the slip surface lay at a shallow depth in the original soil, probably reaching only 1 m below the base of the existing abutment.

As a first measure, the immediate mechanical cause for the slide, the load of the embankment, was reduced by removing the embankment fill down to approximately between height notations 501.0 and 504.0 (Fig. 6-52).

Then the endangered superstructure of the bridge of prestressed concrete had to be protected as quickly as possible. There was no time to carry out extensive soil explorations and analyses regarding the slide mechanism and the position of the slide joints; therefore activities were limited to designing a sufficiently dependable stabilization on the basis of the available data.

A new embankment fill was out of the question because a renewed loading of soil masses that had already moved once would certainly trigger the next slide; consequently, the bridge had to be extended. The pier replacing the old abutment was provided with diaphragm-wall panels (220 × 80 cm) as footings which were designed to take up the load of the pier. These footings intersected the slide body, which had meanwhile settled, without further disturbing it, and safely transmitted the load of the construction to load-bearing soil layers (Fig.

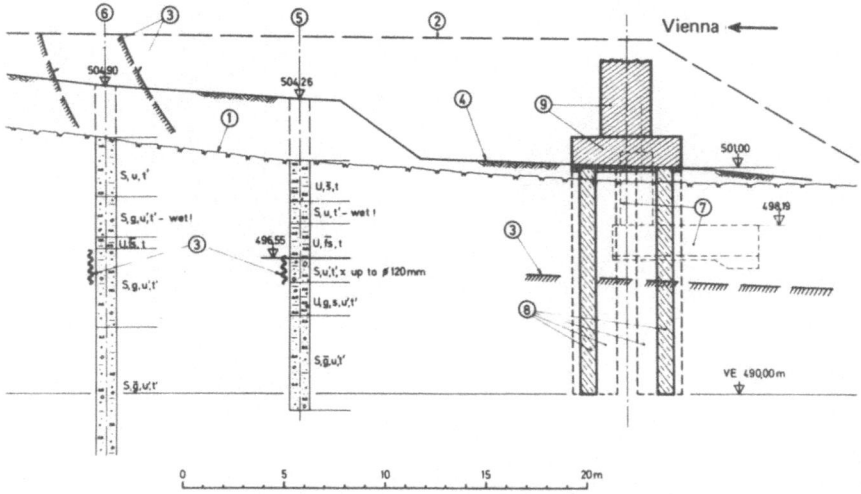

Figure 6-52 Section of slide and foundations of old abutment and new pier; (1) original ground; (2) fill at time of failure; (3) cracks and presumable position of slip surfaces; (4) partially dislocated embankment fill; (5) shaft (excavation was discontinued at 496.55 m on account of uplift); (6) boring; (7) old abutment construction; (8) diaphragm-wall panels; (9) new foundation slab with pier. (For symbols used in soil profiles see 6.6.)

6-52). The diaphragm-wall panels next to the old abutment first served as support for the last bridge span. The panels reached 0.50 to 1.0 m into the gravel layer present at height notation 490.5; at the heads they were connected with a stiff foundation slab (Fig. 6-53) For a safety factor of 2, an admissible vertical load of 2 MN per panel was estimated, thus providing for a sufficient admissible load of 12 MN for the whole pier footing.

The estimates regarding eventual horizontal forces acting upon the group of diaphragm-wall panels were based on the following considerations: In principle no further slide movements were to be anticipated after the removal of the dislocated embankment body, provided water from precipitation was prevented by diversion measures from entering into the soil of the slide area. Secondly, the load of the construction would no longer act upon the slide body. However, for the sake of the safety of this major construction, horizontal forces acting upon the uniformly embedded diaphragm-wall panels were assumed. A parabolic slide body with a 45° gradient towards the axis of the bridge was assumed; Figure 6-54 shows the dimensions. The slip surface, for which no resisting forces were taken into account, was assumed to have a 1:3.5 inclination (vertical to horizontal), this being somewhat steeper than the original ground surface. Since the panels of the footing were connected by a stiff foundation slab, the calculated load was distributed uniformly over all six panels and a sufficient load-bearing capacity could be proved.

Finally, the bridge was extended by two 25-m spans, whereby the additional pier and the new abutment were provided with deep foundations. By these

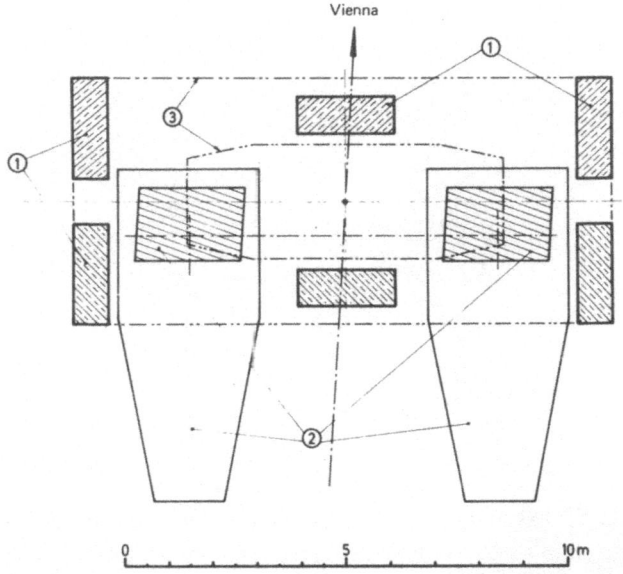

Figure 6-53 Plan of diaphragm-wall panels; (1) diaphragm-wall panels; (2) old abutment construction; (3) profile of new foundation slab or new pier.

measures a high embankment fill was avoided and no further movements occurred.

Perhaps it would have been possible to achieve the stability of the slope by extensive drainage, but this would have required time-consuming construction work. Then it would have been necessary to wait until the soil was sufficiently

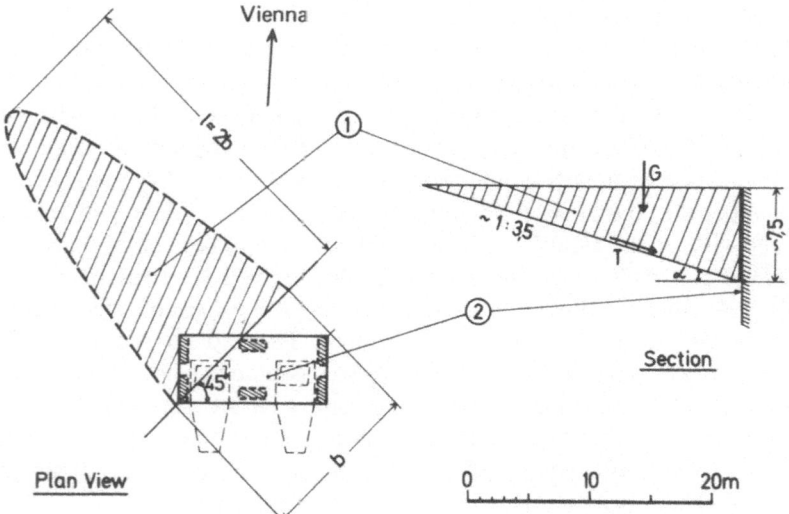

Figure 6-54 Plan and section of model used to estimate horizontal forces acting on group of diaphragm-wall panels; (1) slide body; (2) foundation panels.

drained before construction of the new abutment could be started. As the bridge had to be completed quickly, the chosen solution must be considered the right one.

6.2.2 Steep Cutting Slopes of Loose Soil are (Temporarily) Retained by Shot-Concrete Skins that may be Reinforced and Anchored — Tokyo, Japan, and the Zwenberg Bridge for the Tauern Railway, Carinthia, Austria

Cutting slopes susceptible to slides may be retained with a shot-concrete skin, whereby this measure may either serve a temporary stabilization for the duration of construction work or aim at a permanent stabilization.

Similar to concrete, shot-concrete consists of aggregate with grain sizes in the range of sand to gravel, cement, water, and certain chemical additives, such as agents that accelerate setting; but contrary to concrete, the maximum grain size is in the range of 15 to 25 mm. Shot-concrete is applied by dry-spraying (water is added at the nozzle) or by wet-spraying.

For the purpose of retaining a cutting slope, the shot-concrete skin is at least 10 to 15 cm and at the most 30 cm thick; normally it is reinforced, preferably by steel nets. Currently, a new method is being developed, especially in the FRG, whereby the reinforcement is added to the aggregate in the shape of steel fiber, the so-called steel-fiber concrete. If the soil to be retained is water-bearing, it is absolutely necessary to provide the shot-concrete skin with weep holes; otherwise, dangerously high water pressure will build up.

Retaining cutting slopes with shot-concrete skins offers the special advantage that they can be carried out very quickly. The damaging influences of water and air upon the soil virtually cannot take effect. The method is suited for all types of soil, loose soils with any grain distribution (clay, silt, sand, gravel) and also loosened or decomposing rock. As the shot-concrete hits the soil, a firm bonding or interlocking takes place, so that the slide-prone soil is stabilized at least for the time required to carry out construction work.

This bonding process, which still cannot be fully explained and until today escapes all known static calculations, probably represents a combination of several effects. On the one hand, the loosening of the soil and a corresponding loss of shear strength are prevented and the interlocking–cohesion is maintained and reinforced by the shot-concrete skin. On the other hand, a chemical reaction probably plays a part in the process, e.g., in the reaction between cement and the silicates of the soil, which causes an increase of cohesion and thus of free-standing height.

In special cases cutting slopes retained by shot-concrete skins with a gradient of 40° and more may also be anchored. Here it has proved practical to construct the load-distributing ribs and the anchor beams by the shot-concrete method; thereby, concreting with forms, involving much more work, may be avoided. For longer cuts the construction of berms is recommended.

Figure 6-55 shows the stabilization of the construction pit for a high-rise building that was erected in *Tokyo* right next to a large hotel. The subsoil

Figure 6-55 Lining of construction pit in Tokyo, with reinforced shot-concrete skin.

consisted of relatively stiff, sandy-clayey silt. Holes in the shot-concrete skin took care of the drainage.

Another example is the lining of the construction pit for the southern abutment of the 200-m-wide *Zwenberg Bridge* (Fig. 6-56), which had to be constructed for the purpose of flattening a curve of the Tauern Railway. Since the load-bearing rock required for the foundation of this major construction was 24 m below ground surface, an open construction pit had to be excavated to that depth. The soil consisted in part of loose sand and gravel, and of very soft, fractured, slightly water-bearing rock. If the shot-concrete method had not been developed to such perfection, the construction of expensive retaining walls or even anchored walls would have been necessary.

It goes beyond the scope of this book to deal with all possible applications of the shot-concrete method. To mention a few, shot-concrete may be used to line galleries and shafts, with or without anchoring, and all kinds of containers, and to strengthen weak reinforced-concrete elements, such as beams and piers.

6.2.3   Installation of Stone Wedges at Slope Toe to Increase Friction, Drain and Ballast the Toe—Ybbsitzer Heights, Lower Austria

In the winter of 1965, a part of Federal Highway 253 crossing the Ybbsitzer Heights was realigned. The new route followed more or less the old alignment, but its greater width and smoother alignment required a higher embankment fill. The embankment, up to 6 m high, was indented with the subsoil by steps

Figure 6-56   Lining of the preliminary cut at Zwenberg Bridge, Austria.

and the fill consisted of a well-compacted mixture of sand and gravel, as the scarps of the slide distinctly showed later on (Fig. 6-57).

During the summer of 1966 the first slide occurred in this area. The embankment body slumped by about 0.5 m, and cracks and bulges showed in the downhill zone to a distance of about 50 m, but then movements stopped.

In December of 1966, warm temperatures and rainfall set in while the ground was covered by snow. Helped by this strong inflow of water, a section of the road embankment about 250 m long, together with a downhill part of the slope 200 m wide and overgrown with bushes, slid down.

Shortly after this landslide, ram soundings were carried out downhill of the road to explore soil conditions. There was a surficial layer of brown loam (sandy-clayey silt) with more or less frequent embedments of coarse gravel and stones. The subjacent layer, at depths varying between 1.0 and 4.0 m—owing to the severely fissured ground surface—consisted of blue-grey clay slate. Both layers were relatively impermeable.

Uphill of the old road the slope had a steeper zone with a 45° gradient, probably stemming from the cut for this road. Above this steep zone the mean gradient was between 28° and 30°, corresponding to the general formation of the terrain. These steeper, forested zones showed no signs of old or recent landslides.

Several exploratory trenches were dug uphill of the road reaching to depths of 2 to 3 m. Below the humus of the forest, first brown sandy-clayey silt with single stones was encountered; below a depth of about 1 m there were more stones, still showing the former rock joints, filled with weathered loam; then

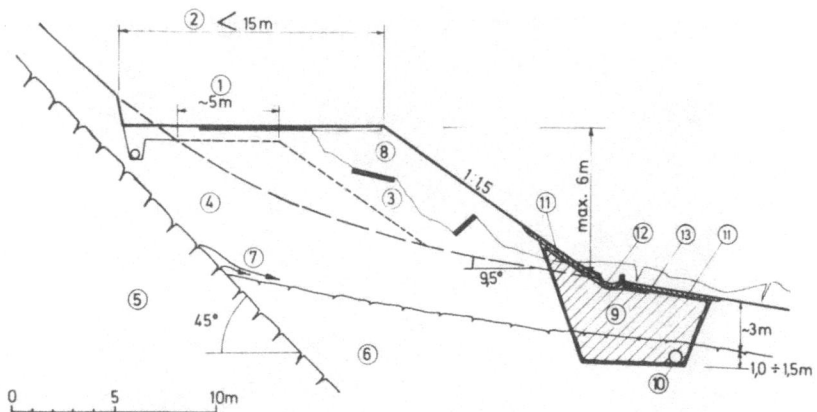

Figure 6-57 Section of dislocated road and stabilization measures; (1) old road; (2) widened road; (3) slide of embankment for wider road; (4) weathered layer; (5) limestone (fissured); (6) clay slate (stiff); (7) crack water; (8) gravel embankment body; (9) gravel and stone fill as embankment toe; (10) drainage; (11) dense soil; (12) open drainage; (13) PVC filter fabric.

followed sound bedrock. The profile of the actual slide area, down to the downhill forest, had an average gradient of only 9.5°.

An examination of the slide body itself revealed an uncommonly severe fissuring, with deep cracks, slumps, and tilted soil blocks. The soil was very moist, as evidenced by water outlets and many new small pools.

The severe fissuring pointed to a relatively shallow slip surface which was apparently identical to the surface of the clay slate encountered at depths between 1 and 4 m. This meant that the clay slate extended parallel to the slope surface and was present only in the flatter part of the slope. Approximately where the old road had been, the bedrock dipped with a steep gradient below the clay slate; therefore, the slope uphill of the road, in spite of being very steep, did not show any cracks or slides.

The heavy soaking of the flat zone of the slope, as evidenced by springs and pools, indicated that the development of a lubricating layer at the zone of contact between weathered sandy-clayey silt and clay slate was the cause of the slide. This also explained how a slide movement could occur at a gradient of only 9.5°, even though the sandy-clayey silt as well as the clay slate certainly had an angle of friction above 9°.

Immediate measures were proposed for the stabilization of the road embankment because in the presence of the deep fissures and scarps any further precipitation would have triggered new slide movements. These measures mainly consisted of:

- Excavation of a trench, parallel to the road on the uphill side of the embankment, reaching about 1 m into the solid clay slate; the excavated material was dumped outside the slide area.

- Installation of a large drainage pipe (diameter—200 mm) at the bottom of the trench, with a high gradient, parallel to the road. It was not considered safe to divert the water along the line of the dip through the slide area, since the creep movement of the slide might have destroyed the drainage lines.
- Backfilling the trench with gravel and riprap so that it would serve as support for the embankment toe. This backfilll had the double purpose of increasing friction and draining. With today's technical standards this supporting body would have been protected against clogging by wrapping it in filter fabric (Fig. 6-35), thus ensuring a lasting effectiveness. In addition, it is advisable to divert surface waters with open channels to avoid an overburdening of the trench. These supplementary measures are also shown in Fig. 6-57.
- Surface drainage of the pools that had developed with the slide, and filling the cracks with mortar.

After the above measures had been completed, slope stability was monitored by registering horizontal and vertical movements as well as the water discharge from the springs downhill of the slide area and from the concentrated water outlets in the slide slope itself. The stabilization measures proved effective and no further movements of the embankment were registered.

6.2.4   Soil Exchange at Embankment Base—Embankment Slide near Oberpullendorf, Burgenland, Austria

In the spring of 1969 the high embankment body between km 50.4 and km 50.5 of the railway line Oberpullendorf-Rattersdorf started to settle, making it necessary to tamp the tracks again and again; the causes were at first not discernible (Fig. 6-58).

The shape of the scarps and the presence of numerous pools downhill of the embankment led to the conclusion that the causes were purely mechanical. This seemed to be confirmed by the sharp edge of the scarp on the uphill embankment flank and the slumping of several culvert rings (Fig. 6-59).

In order to explore soil conditions in more detail, soil profiles I and II were made. Furthermore, a number of ram soundings were carried out. These showed a multilayered structure of the subsoil of the Pannon, with alternating layers of sandy-clayey silt and sand. A 6-m-deep exploratory shaft (Fig. 6-59) completed the picture. During an inspection it was noted that the shaft had filled with water flowing amply from two sand horizons. In a trench excavated at the same level, a considerable flow of water from one sand layer set in at a depth of about 3 m. The quick collapse of the unbraced trench indicated a low cohesion in the sandy-clayey silt layer. A detailed examination of three transverse profiles, taken at the same place but at three different points in time, and the soil analyses, as well as the position of the pools, led to the recognition that the following factors were most probably responsible for the landslide.

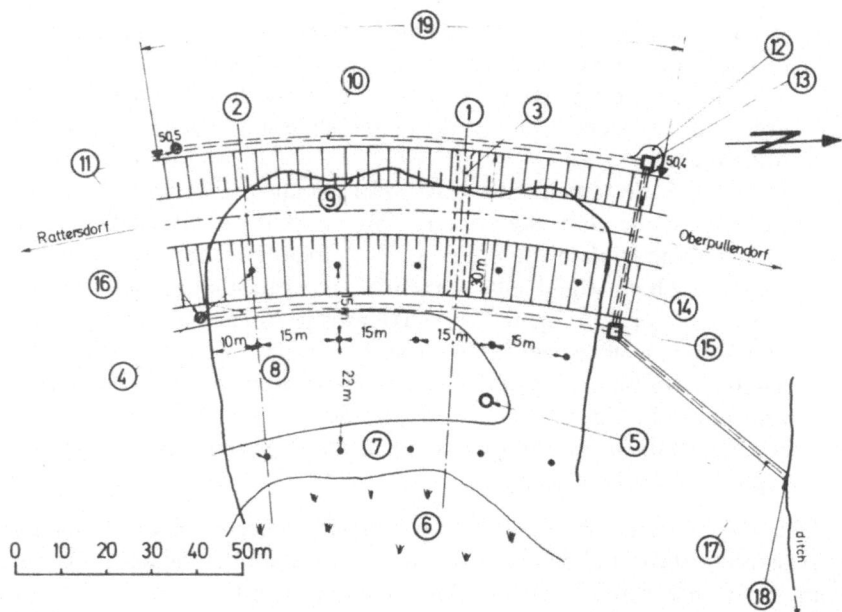

Figure 6-58 Site plan of slide with designed stabilization measures; (1) profile I (see also Fig. 6-59); (2) profile II (see also Fig. 6-60); (3) culvert, diameter—2 m; (4) ram soundings; (5) test shaft, 6 m deep; (6) swampy holes; (7) meadow; (8) field; (9) scarp; (10) uphill subsurface drainage, diameter—200 mm, and surface drainage with open concrete channels; (11) precast flushing shafts; (12) area for collection of surface waters; (13) collection shaft for surface waters, with grating; (14) precast concrete culvert, diameter—0.9 m; (15) collecting pit with entry; (16) downhill subsurface drainage, diameter 200 mm; (17) precast concrete pipes, diameter 0.90 m; (18) discharge into ditch; (19) section of embankment to be exchanged.

Down to a depth of 6 m at least two sand horizons were water-bearing (Fig. 6-60); these horizons were wedging out farther downhill, in the flat zone of the slope. This flow of water eroded the completely cohesionless sand (size of grain—0.06 to 0.3 mm), and the sandy-clayey silt layer above lost its support and caved in. This was the cause for the many old scarps around the pools as well as the pools themselves, since crumbling sandy-clayey silt layers prevented the water from seeping away. This in turn casued a backwash and an increase of the pore-water pressure, and explained the high water level in the exploratory shaft in spite of the steep gradient of the sand horizons (see Case History 6.1.4.1 (II).

Regarding the embankment itself (grain distribution: d_{10} = 0.1 mm; d_{60} = 1.0 mm), it certainly represented a major load exerting an additional driving force upon the sandy-clayey silt layer with a gradient of only 10°. In the course of time this layer, which was cut at the toe by subsequent failures, apparently reacted with minor downhill creep movements, which had increased by the spring of 1969 to the degree that the embankment, robbed of its base, reacted

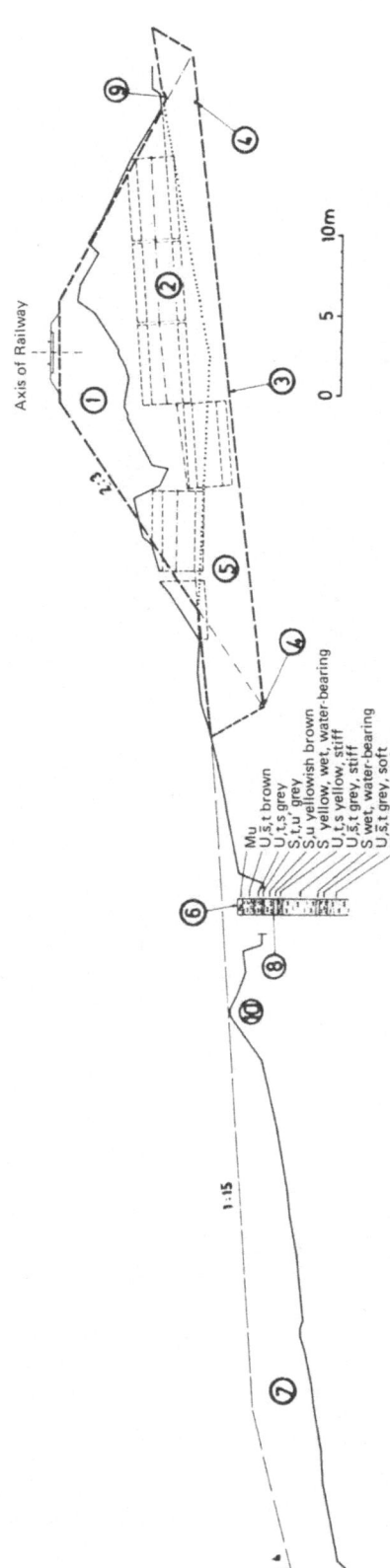

Axis of Railway

Mu
U,š,t brown
U,t,s grey
S,t,u grey
S,u yellowish brown
S yellow, wet, water-bearing
U,t,s yellow, stiff
U,š,t grey, stiff
S wet, water-bearing
U,š,t grey, soft
U,t,s grey, soft

0   5   10m

Figure 6-59 Profile I, at km 50.429, with designed stabilization measures; (1) embankment fill of basalt tuff; (2) concrete culvert, diameter—2 m; (3) depth of excavation for embankment base (layer of sand and gravel); (4) drainage, diameter—200 mm (slotted plastic pipes); (5) gravel for embankment base; (6) test shaft; (7) design of fill, using excavated material; (8) level of new embankment base; (9) open concrete channel; (10) bulge due to outflow of sand. (For explanation of symbols used in soil profile, see 6.6.)

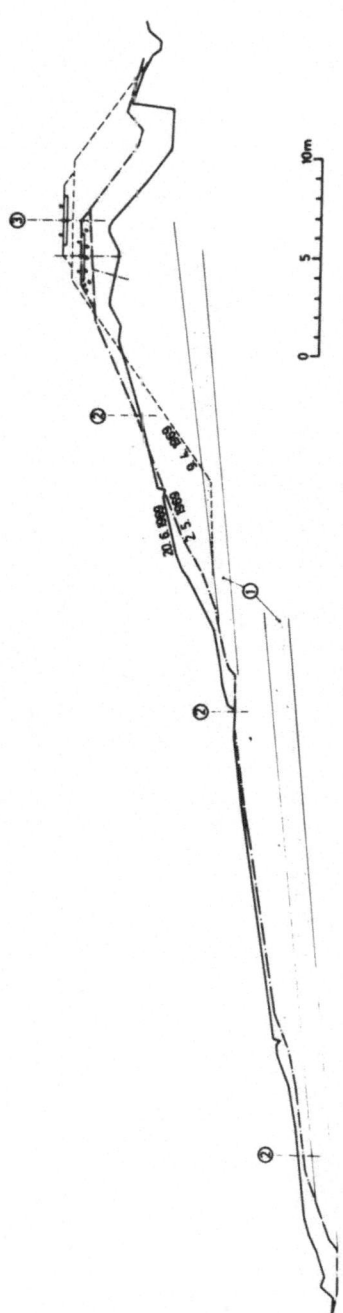

Figure 6-60 Profile II, at km 50.470; (1) sand lenses; (2) row of ram-soundings; (3) axis of embankment.

with a slide. The embankment fill, which was stiffer than the original soil of the slope, was not able to take up the large creep deformation without failure, that is, without a marked slip surface.

When using the transverse profiles to compare the individual stages, it became evident that a strictly translational movement had taken place in the original soil (starting at the pools and moving uphill), while—as a consequence—the embankment itself reacted with a slide. The downhill bulges and the uphill slumps were further evidence of this process. The water coming from the leaking culvert at km 50.429 may have speeded up the process.

For the stabilization of the railway route the following measures were proposed:

- Removal of the embankment destroyed by the slide and excavation of the base course down to the sand layer, since laboratory tests had indicated that the layer above was not load-bearing.
- The width of embankment should be determined by the extension of the embankment flanks at a 2:3 inclination (vertical to horizontal). The sidewall of the construction pit should have a maximum gradient of 60°. Excavation work should be carried out by sections, each 20 m long, so as not to undercut the uphill slope for too long a period of time.
- The construction pit should be drained continuously. The water collected should be diverted to the natural ditch downhill of the exploratory shaft. The trench should be excavated prior to excavating the base course.
- In order to provide the embankment with a firm and at the same time permeable base, the excavation should be first backfilled with well-compacted gravel.
- The embankment body should be filled with a 2:3 inclination, using basalt tuff from the quarry near Oberpullendorf, as was used for the old embankment. This material may be susceptible to weathering by the sun and is easily crushed (upon delivery 0.5% 0.1 mm; after tamping in the Proctor test 49% 0.1 mm), but may be used as embankment fill if the base consists of good gravel. Proper compacting during filling operations and periodic controls are indispensable.
- The completed embankment should be sown with grass as soon as possible.
- A careful drainage of base and surface should be carried out (Figs.˙6-58 and 6-59). Open concrete channels on the uphill side of the embankment, joining in a collection shaft, were recommended for the surface waters. The base course was to be drained with slotted PVC-tubes, also discharging into the collection shaft, from which a culvert should lead to the downhill collection shaft. A downhill longitudinal drainage starting at the deepest point of the base course excavation was to join that same shaft, from which a drainage pipe was to lead to the natural ditch to the southeast, which was to be paved until the end of the embankment fill.
- The material from the old embankment should be dumped downhill to serve as ballast for the embankment toe, starting at a distance of about

25 m from the axis of the railway (Fig. 6-59). The ballasting fill should have at the head a 1:15 and at the foot 1:4 (vertical to horizontal) inclination. Before filling the ballasting body, to be carried out in thin layers, humus and bog should be removed to a depth of about 0.5 m, whereby the water pools should be carefully avoided. For compacting, the weight of the bulldozer is sufficient. Under the circumstances, it is advisable to carry out the ballasting of the foot in the manner of a Terzaghi filter, especially where sand layers are wedging out.

For cost reasons this project was not carried out especially since there was very little traffic on this railway route leading to the Hungarian border. Instead, a bus line was put in operation.

Again and again it may be observed that embankments that have remained relatively stable since their construction, often for 90 years and longer, will suddenly slide as a whole or in part, because at the time of their construction soil-mechanical considerations and analyses, which are a matter of course today, were not taken into account. Neither the load-bearing capacity of the subsoil nor hydraulic conditions (drainage requirements, etc.) were investigated, nor were filling operations carried out according to the rules that are valid today, rules that call for a certain density, a certain unit weight, a certain water content and a certain E-value; at that time geotechnical engineering had not advanced to that standard. Actually, it is a marvel that so many old embankments have served over such a long time.

An incipient embankment failure usually announces itself by major cracks and an embankment deformation. It is the duty of the observant railway superintendent to report such occurrences immediately to his superior, who must then initiate appropriate measures.

If such changes develop only slowly over longer periods of time, they must be monitored by periodic horizontal and vertical measurements from benchmarks that are registered graphically. As soon as a marked acceleration of the movements is observed, nearing a point known from experience as being critical, traffic on that route has to be stopped (see 5.1.7).

6.2.5 Lining the Bottom and Sides of a Cut with Gravel—Tailrace Channel, Silz Power Plant, Tirol, Austria

From the Silz Power Plant to the Inn River, the approximately 1.1-km-long tailrace channel cuts through a zone consisting of very loose, sandy-gravely alluvions—deposits of the Inn River. It was decided not to line this channel (water passage—48 m³/sec; flow velocity—about 1 m/sec) with concrete. Starting at the concrete discharge gate (station 0.0), from station 2.68 to station 23.00, the channel was lined with a layer of riprap approximately 0.7-m-thick on top of a 0.15-m-thick layer of gravel (grain diameter 32 to 64 mm); up to station 10.0 filter fabric was added. From station 2.68 to station 5.0 the layer of riprap was increased along the joints between the bottom and sides to a height of 1.0 m (inclination of sides between 1:1.5 and 1:1.2—vertical to horizontal). Above water level the bank was seeded (Fig. 6-61a).

Starting at station 23.0, the bottom and that part of the sides that was under water were covered with gravel (grain diameter < 64 mm). The thickness of this layer was 0.5 m at the bottom, 1.0 meter along the joints between bottom and sides, and then tapered off to 0.6 m at the water level. Here the inclination was kept at 1:3 from the bottom to a height of 1.2 m, after which the inclination was 1:2 (Fig. 6-61b).

The basic idea behind this design was not to seal off the channel against the adjacent very permeable soil, as would have been the case with concrete slabs, but to allow a free play between the water flowing in the channel and the ground water in the soil. A possible flow of ground water into the channel would be of no significance and in turn, a possible seepage of channel water into the ground water would be quite harmless. The water flowing in the channel comes from the Finstertal and Längental reservoirs, which hold very clean water from the mountains.

By this most economical method of lining a channel, which is sufficiently smooth and also ensures the stability of the sides of the cut, an expensive concrete lining is avoided. A concrete lining only makes sense if the channel has to be periodically cleaned of sludge; however, this was hardly necessary because the water was so clean.

The channel for the Imst Power Plant, situated upstream, which has been in operation since 1956, has a quite similar profile. In 1977 the profile was inspected and it was found that virtually no changes had taken place during the 22 years since its construction.

6.2.6 Installation of Terzaghi Filter to Check Erosion of Easily Moving Layers of Fine Sand

No words need be lost regarding the construction and action of a Terzaghi filter (Terzaghi and Peck, 1967), but based on my practical experiences, I feel it should be pointed out that in the case of spontaneous hydraulic ground failures, which mostly are very dangerous, time pressure usually makes it impossible to install the Terzaghi filter in several layers, as prescribed. When such catastrophes occur, the immediate danger can often be averted by spreading uniformly graded aggregate, which is usually at hand, over the endangered spot. This aggregate should not contain grains with a diameter less than 0.1 mm.

By this measure the particularly labile grains of fine sand and coarse silt (diameter—0.2 to 0.02 mm) are held back; the smaller grain fractions normally have so much cohesion that they are not set in motion during a hydraulic failure (see 6.2.4 and Case History 6.1.4.2 (II).

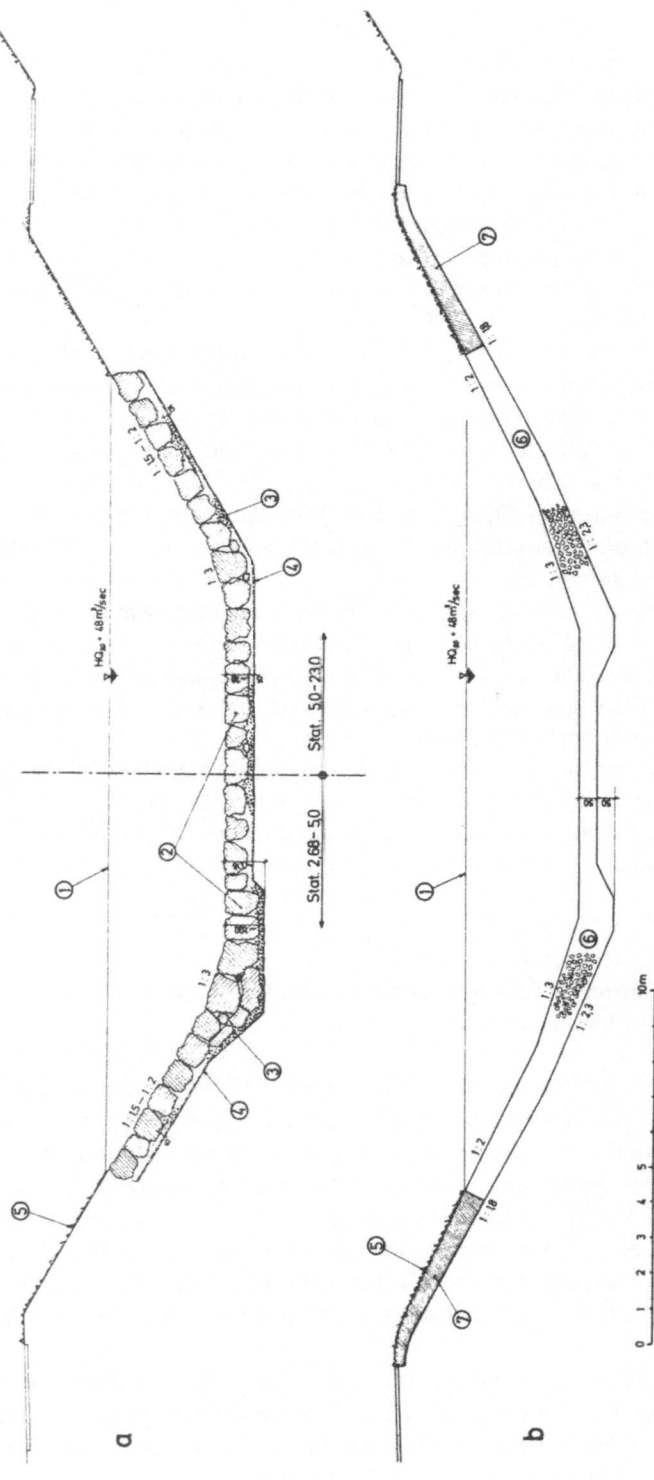

Figure 6-61 Lining of tailrace channel Silz; (1) highest water level; (2) riprap; (3) oversized grain, diameter—32/64 mm; (4) filter fabric; (5) seeded area; (6) stones, diameter > 64 mm; (7) soil deposits.

6.2.7 Installation of Stone Ribs or Stabilized Soil Material Parallel to Slope (this method may be used only if the slide circle is intersected)—Brick–Clay Pit, Budapest, Hungary (Kézdi, 1976)

Soil conditions at the brick–clay pit near Budapest were marked by a subjacent layer of stiff clay (local term—*Kiscella-Ton*) with a 1:15 inclination (vertical to horizontal), on which rested slope-parallel, weathered, less-stiff layers of yellow and brown clay ($w = 0.20$; $w_p = 0.25$; $w_L = 0.55$).

Slides set in caused by the working of the pit at a cutting inclination of 2:3 and steeper. It could be assumed that the slip surface was at a depth of about 8.0 m in the upper yellow and brown layers. The layers above the slip surface were creeping; their structure was weakened by weathering and they had a low shear resistance, which was further decreased by flow pressure and pore-water pressure.

A sufficient stabilization of the slope could only be ensured if the groundwater table was lowered by means of seepage trenches. However, as this still would not lead to complete drainage, the slope also had to be supported by the seepage trenches following the line of the dip which were backfilled with coarse gravel wrapped in filter fabric containing a plastic filter tube. As this method is only effective if the slide circle is intersected, the trenches must reach below the slip surface. Their effectiveness must be monitored by registering eventual further slide movements and water discharge (3.3.1)

6.2.8 Ballasting Fills as Support for Embankment Toes

6.2.8 (I)   Freeway, German Federal Republic
Gottstein (1936) reports about an embankment fill for a freeway (mean inclination—2:3 (vertical to horizontal); width of base—51 m) that rests on a 45-m-thick layer of silt (marine deposits of the ice age). In the zone of the embankment toe the silt was covered by a 1.5-to-4.0-m-high layer of peat.

The embankment was designed with a height of 8.7 m, but when the fill had reached a height of 6.0 m, almost half of the embankment body slid down on a circular-cylindrical slip surface. This embankment failure extended over a length of 120 m and within a few days the block slumped by 3 m. Simultaneously, the ground surface, at a distance of about 10 m, was raised by about 2 m. Another characteristic symptom was the 1:12 inclination of the crown of the slumped embankment to the axis of the road.

Near the surface, the silt had the following soil coefficients: $w \cong 50\%$; $w_L = 65\%$; $I_p = 23\%$; density $\gamma_s = 2.65$ t/m^3; $k_f = 6 \cdot 10^{-7}$ cm/sec; grain size: sand 54%, silt 46%; 80% of the material soluble in HCl. A recalculation of stability after Petterson and Terzaghi (Terzaghi, 1936a) yielded a safety factor $\eta = 1.0$, confirming the limiting equilibrium of the embankment. The embankment was stabilized by the following measures:

- A ballasting fill 2.5 m high and 40 m wide consisting of material as heavy as possible was dumped on the peat layer.

- The remaining 2.7 m of the embankment were filled with the lightest possible material (blast furnace slag).

These measures increased the safety coefficient to $\eta = 1.6$.

Regarding the sensitivity of this construction method, it is interesting to note that where, by mistake, the ballasting fill was only 1.5 m high, and the remaining 2.7 m of the embankment were filled with heavy gravel, the same type of slide set in immediately.

6.2.8 (II)   Vermont, United States (Kézdi, 1976)

On August 18, 1970, during construction work, a ground failure occurred below the embankment of the four-lane freeway between Fairhaven and Castleton, Vermont. The subsoil consisted of soft, varved clay and silt of the post-glacial period. At a depth of about 25 m below the original surface, bedrock of densely layered slate was encountered, which was forming, together with the embankment body for the second roadway, a closed trough under the embankment; this prevented the decrease of pore-water pressure in the soft layers.

The soil had the following coefficients: Clay: density $\gamma = 1.89$ t/m^3; $w = 0.42$; $w_L = 0.37$; $I_p = 0.16$; $c_{cu} = 0.02$ MN/m^2; $\varphi_{cu} = 12°$; $q_u = 0.037$ MN/m^2. Silt: $w_L = 0.3$; $I_p = 0$.

At the time of failure the embankment fill had reached a height of 12 m (the designed height was 13 m); the lateral berm was about 5 m high and 25 m wide. Embankment and berm sides had a 1:2 inclination (vertical to horizontal).

During filling operations, a settling by about 50 cm and a pore-water pressure of 1.4 bar were registered. Calculation with only two slide circles had yielded a safety coefficient $\eta = 1.17$. During failure, which occurred with a velocity of 30 cm/min, the embankment slumped by 3.0 to 3.5 m and the berm suffered a lateral dislocation by 8 to 10 m. Shear strength after failure, measured with a vane apparatus, amounted to 0.043 MN/m^2. A recalculation on the basis of this value resulted in a safety coefficient $\eta = 0.989$.

To reconstruct the embankment, first berms were constructed on both sides. These were 65 m wide (more than twice as wide as the old berm) and 8 m high (60% higher). Filling operations for the embankment were taken up only after the berms had been completed; for both, the previous 1:2 inclination was retained. Since 1976 this part of the freeway has carried traffic without any trouble.

What lessons may be drawn from this ground failure?

(1)   The stability analysis was based on only two soil-mechanical experiments; this is not enough in such a critical case.

(2)   The safety coefficent before failure was calculated with a very simplified computer method. The resulting safety coefficient was close to 1; in this case calculations must be extended over several slide circles.

(3)   The first layer of the embankment fill should have consisted of filtering material to a height of about 1 m; this would have speeded up the consolidation process.

(4)   The safety coefficient was decreased, since the projected density of the embankment fill of γ = 20.8 kN/m³ was increased to 22.4 kN/m³ by rolling rock into the clay.

(5)   Finally, pore-water pressure was increased considerably because below the critical zone the bedrock and the neighboring embankment being filled formed a trough, which constricted the flow of water from the clay. Here vertical drainage would have been advantageous.

6.2.9   Retaining Walls

6.2.9.1   *Gravity or Cantilever Retaining Walls (see 3.2.3)*

6.2.9.2   *Bore-Pile or Diaphragm Walls (for example with T-shaped panels)*
without Anchoring—Construction of the Olympic Road, Rome, Italy
In the construction of retaining walls, bore-pile and diaphragm walls have taken the place of the traditional construction elements. This new construction method permits more modern design at equal cost; this is by far a better adaptation to the individual requirements and an optimal utilization of the available space. Previously, the construction of retaining walls, regardless of whether they were massive or of the cantilever type, first required an open braced excavation to the foreseen depth and only then could the wall be erected; in most cases the construction pit had to be carefully strutted and water conditions had to be dealt with.

The new type of wall can be carried out from the ground surface and already represents the completed construction, so that no further work is required after the removal of the soil in front of the wall. The static concept of this wall type is to achieve bending resistance with the use of T-ribs; in fact, when placed towards the terrain, the wall has the shape of a T-beam and acts like a cantilever slab. A horizontal slab on top of the ribs (relief slab; Fig. 6-62) reduces the bending moment and in addition screens the wall against a major part of the earth pressure. It is also possible to stabilize the retaining wall further by a slab fixed on one side into the retaining wall and forming the roadway. This is demonstrated by the following case history dealing with the retaining wall below the monastery Padri Bianchi, which was carried out in connection with the construction of the Olympic Road in Rome in 1961 (Figs. 6-62 and 6-63).

The construction of this freeway required a deep cut in front of the old monastery on a pile foundation; it was not possible to anchor the retaining wall. A diaphragm wall with ribs was constructed. In plan it is a T-beam and the exposed face is a plane; at a depth of approximately 4 m a reinforced concrete slab was installed on top of the ribs to reduce the bending moment. The roadway consists of a concrete slab on cross beams fixed into the retaining wall. The retaining wall, reaching up to 11 m into the natural soil and having an exposed face up to 14 m high, permitted a much better utilization of space as compared to traditional types of retaining walls.

Basically, what has been said in 3.2.3 is also valid here, but it should be pointed out that the concept described above also offers advantages for the

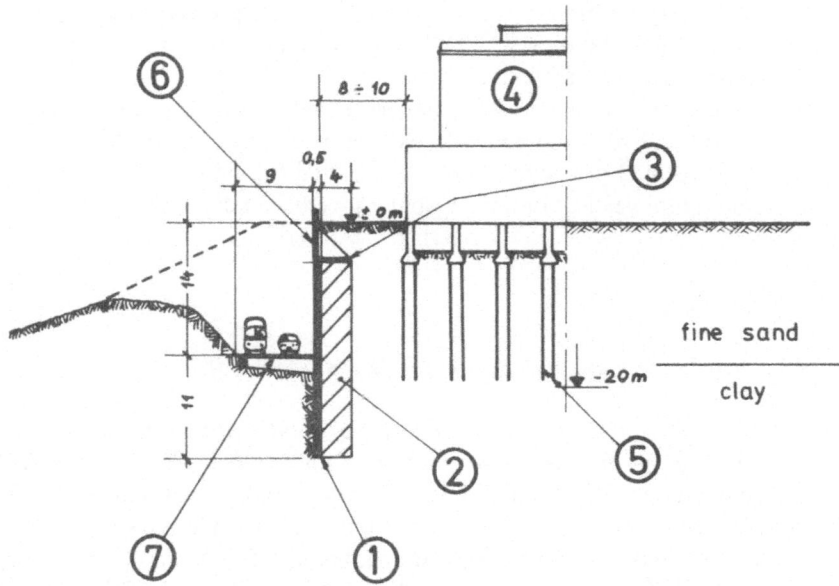

Figure 6-62 Section of retaining construction; (1) bentonite diaphragm wall; (2) ribs; (3) relief slab; (4) building; (5) existing pile foundation; (6) retaining wall concreted on top of diaphragm wall; (7) roadway slab tensed into retaining wall.

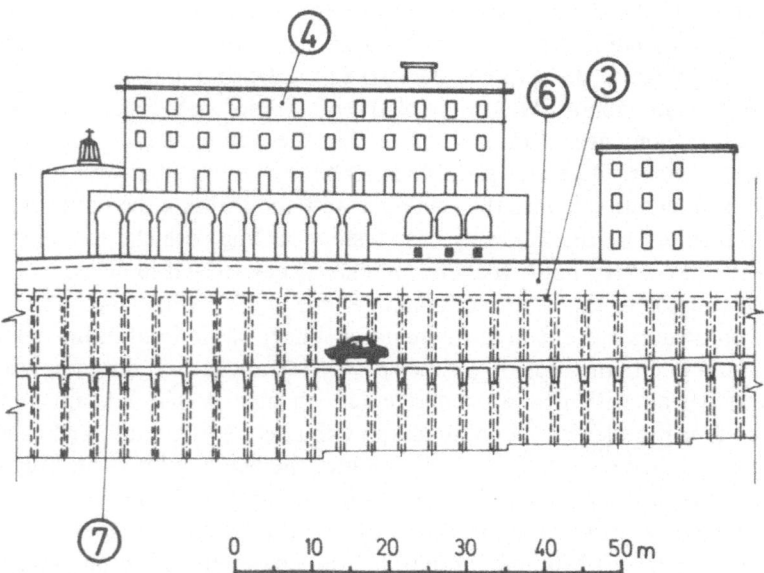

Figure 6-63 View of retaining construction; (3) relief slab; (4) building; (6) retaining wall on top of diaphragm wall; (7) roadway slab tensed into retaining wall.

construction of bridge abutments. For example, where an overpass and under-pass for a freeway crossing are to be constructed, the following construction method will in most cases save time and money:

(1) Construction of the abutments and any piers that may be required from the ground surface with diaphragm-wall panels; an arrangement should be chosen that is best suited for the purpose and local conditions, i.e., *T*-, *H*- or *I*-shape.

(2) Construction of the superstructure of the bridge resting on the diaphragm-wall panels, whereby the original ground surface supports the forms for concreting.

(3) Excavation of the underpass, after the superstructure has set, whereby the bridge may serve as strutting between the two abutments.

As compared to traditional construction methods, the following three operations need not be carried out for the construction of the underpass:

• Excavation, bracing of the construction pit for the abutments, and building of the forms
• Construction of forms for the piers
• Erection of a scaffold for the construction of the superstructure.

The diaphragm-wall method offers itself especially for deep foundations in weak soil. Under such conditions the diaphragm wall takes over in a single construction operation the double function of retaining construction and foundation element (ICOS 3, 1968).

6.2.9.3 *Anchored Walls of Load-Bearing and Non-Load-Bearing Bore-Pile or Diaphragm-Wall Panels*

The construction of freeways through mountainous territory often makes it necessary to cut into high and very steep slopes. In the following, two case histories from the construction of the Tauern Freeway in the Austrian province of Salzburg are presented (Brandl, 1976b). The soil on these construction sites consisted of silty talus and weathered talc and mica slate, respectively, and the slopes were, in their original state, near or at limit equilibrium.

Example I: concerns a section of freeway about 800 m long crossing a mostly very steep and wet over 1,000 m high slope; up to 45-m-high cuts and 25-m-high narrow slope wedges were required. The bedrock consisted of mica slate, with a high share of mica; its bedding planes were more or less parallel to the ground surface; consequently, there was danger of a bedding-plane slide, especially at the undercuts. On top of the bedrock rested, in part, huge masses of decomposed rock and deposits from debris avalanches and moraines, which were heavily soaked.

In the course of cutting operations, extensive slides set in over a length of about 250 m, with deep cracks and up to 3.5-m-high staggered scarps, which progressed uphill and finally reached about 450 m into the uphill slope. After a period of rest during winter, movements increased again in spring to several centimeters per week and the cutting slope was pressed out by up to 4 m. Surface drainage trenches did not have a significant stabilizing effect.

In spite of its scaly-platy grain, the talus had a friction angle $\varphi'_r = 25°$ to 30°. Investigations revealed that the slide was not so much due to a decrease of the friction angle caused by cutting operations, but primarily to a flow pressure of seepage water acting parallel to the slope surface and in the direction of the slope face.

As a *first stabilization measure*, 13 heavy supporting blocks (width, 6–7 m; height, 25–35 m; thickness, about 5 m; inclination, 2:3 to 4:7, vertical to horizontal) were installed over 65 m of the slope. Furthermore, 12 drainage borings about 30–40 m deep (rotation core borings; 118-mm-diameter casing tubes with continuously slotted filter tubes) were dug in the area of the roadway, at a rising angle of 15° (Fig. 6-64). The first discharge amounted to 1.2 liters/sec per drainage tube; it gradually decreased to 1.0 liter/sec. Movements of the 31 uphill benchmarks had to be kept under constant observation. These showed that the first stabilization measures had led to a noticeable quieting of the slope but that creep movements of several centimeters per month continued.

Therefore an anchored bore-pile wall, fixed into the rock, was constructed at the toe of the slope in the *second phase of stabilization work*. Supplementary borings were sunk vertically to explore the rock surface and water conditions. In one spot artesian water was encountered, which spurted 2.50 m above ground surface with a discharge of 1.5 liter/sec, later decreasing to 1.0 liter/sec.

The bore-pile wall consisted in the most exposed zone of panels (center distance 2.30 m) formed by two overlapping bore piles with a 90-cm diameter (Fig. 6-65); the piles facing towards the roadway were reinforced. These piles

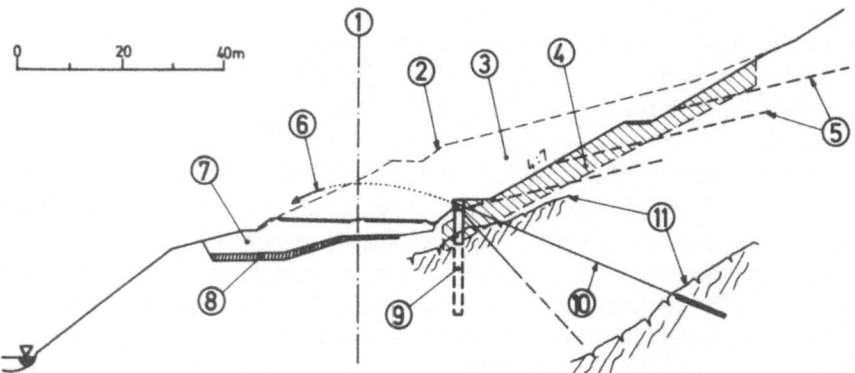

Figure 6-64 Section of slide area and stabilization measures in less critical zone; (1) axis of motorway; (2) original ground; (3) wet soil; (4) supporting ribs; (5) drainage borings; (6) artesian waters; (7) fill; (8) coarse filter; (9) bore piles; (10) rock anchors; (11) bedrock surface.

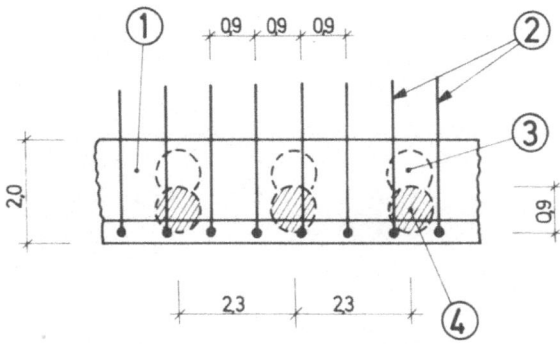

Figure 6-65   Plan of bore-pile wall carried out in most critical zone; (1) head beam on piles; (2) anchor; (3) unreinforced pile; (4) reinforced pile.

have a high resistance moment and the water from the slope can pass freely through the spaces between them. Depending upon quality and position of the bedrock, the bore piles were between 7 and 25 m long.

In the less-exposed zones, walls of single bore piles (diameter—90 cm; center distance—2.10 m) or gravity retaining walls, i.e., "Krainer" walls (6.2.9.5 (I), were constructed. The piles were rigidly connected with head beams so that they could take up and transmit the anchor stresses. Then 90 pieces of 1,200-kN rock anchors were installed at distances between 0.9 and 2.3 m; their inclination varied between 25° and 45° in order to avoid a concentration of strain upon the mountainside. The anchors were between 20 and 53 m long, with an adhesion zone about 8 m long. They were tensed to only 40% of their capacity to allow minor movements.

In the course of pile-boring work, artesian water was encountered in the middle part of the slope, with a multiple discharge as compared to the volume already collected by the drainage borings. Boring work was badly impeded by this outflow of water. During the following anchoring work, jets of water spurted in some places from the casing tubes to distances of up to 30 m. The artesian pressure of 7 bar was much higher than the operating pressure of the boring tools; therefore, relief borings had to be advanced in a fanlike pattern around the anchor borings. The artesian water came from the layer of weathered rock on top of the bedrock surface, which was covered by the dense, silty moraine (Fig. 6-66). More 20-to-30-m-deep drainage borings (diameter—70 mm) were dug to bring down the water pressure in this layer and wells were sunk uphill of the bore-pile wall.

In addition, extensive surface and subsurface drainage was constructed, covering the area uphill of the cut to a height of up to 500 m. The rate of water collected by all drainages was about 200 liter/sec.

Example II: Here the slope was situated in a wide geological fault. To quite some depth the soil consisted of talcy, finely disintegrated weathering products of grauwacke, with occasional boulders and even some embedded hardened

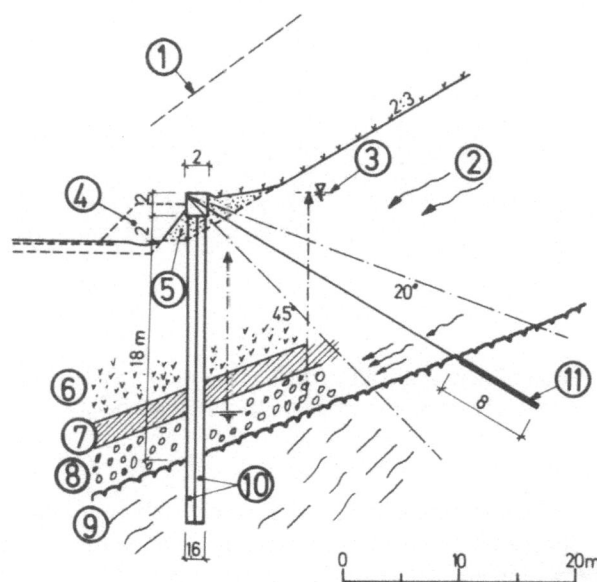

Figure 6-66 Section of anchored bore-pile wall carried out in most critical zone; (1) original ground; (2) slope waters; (3) artesian water; (4) berm; (5) downhill fill; (6) talus; (7) moraine (dense); (8) weathered rock; (9) bedrock; (10) bore piles; (11) rock anchors.

veins. At the slope toe alluviated finely-grained weathered material containing organic admixtures had interlocked with the flood deposits on the river bottom. This soil composition reached to depths of up to 20 m in places; subjacent was a layer of medium to very dense deposits of valley gravel and redeposited gravel moraines. The whole slope was heavily soaked.

Laboratory tests showed that the weathered grauwacke had a very low friction angle and had a tendency to decrease to residual strength $\varphi'_r = 10°-15°$, especially under localized overloadings and under major thrust deformations. Consequently, it had to be anticipated that progressive failure will occur owing to an alignment of the thin, micaceous, finely grained particles along slip surfaces.

Through this slope, which was near limit equilibrium before construction work started, a cut had to be made at a height of about 50 m. First extensive stabilization measures were carried out, such as surface and subsurface drainage, grading down the slope, coffer dams, stone ribs, as well as soil exchanges and ballasting fills.

Although the slope was graded down from an original 3:4 inclination (vertical to horizontal) to a designed inclination of 1:2.5, staggered cracks and scarps several meters high developed, and strong horizontal movements set in, which in part even reached into the uphill forest, causing the downhill zone of the slope to bulge out by almost 4 m and in one place pushing up the roadway base by almost 1.5 m (Fig. 6-67).

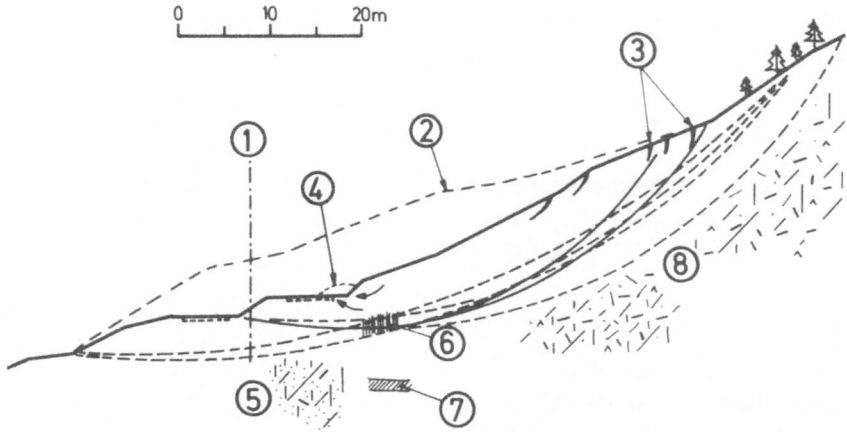

Figure 6-67 Section of slide area; (1) axis of motorway; (2) original ground; (3) cracks; (4) bulge; (5) deposits on valley bottom (organic admixtures, weathering products); (6) soft zones; (7) old slickensides; (8) fine, scaly, talcy weathering products.

As an immediate measure, 17 additional drainage borings reaching to depths between 20 and 35 m were dug at a rising angle of 15° at various heights in the catchment area of the already existing parallel subsurface drainages (total discharge—1000 liters/hr).

Then a 112-m-long anchored retaining wall was constructed, consisting of panels formed by two touching, reinforced-concrete bore piles (center distances—2.80 m; diameters: uphill pile —1.50 m, downhill pile —1.30 m). The panels were connected by a continuous reinforced-concrete head beam to achieve bending resistance, and then the superimposed retaining wall was constructed. Each bore-pile panel was provided with two 1,000-kN alluvial anchors; these were only tensed to 800 kN to allow for movements and to have a strength reserve.

The design of piles and anchors was first based on the slide circles shown in Fig. 6-67. The position of the circles, in turn, was determined by position and shape of the uphill staggered cracks and scarps, the soft zone about 7 m below the roadway base, and the measurements of slope movements. To determine the critical values, the friction angle φ' was varied between 15° and 25°, cohesion c' between 0 and 20 kN/m², and finally, the slide circle drawn in Fig. 6-67 with a full line was accepted as being most critical.

Safety was calculated after the Swedish lamella method, for which a computer program was available. It was assumed ideally that within the slide wedge the opposing earth pressures upon the lamella sides are equal (only moment condition is fulfilled). To calculate anchor strength, the approximate intersection of the resulting forces with the chord of the slide circle (not the slide circle itself) was used. This assumption, which is certainly admissible in the presence of low cohesion, offers the advantage that the resulting value for anchor strength is independent of the height of its application. From the safety coefficient

$$\text{present } \eta = \frac{\Sigma(\Delta G \cdot \cos \alpha \cdot \text{tg}\varphi' + c' \cdot \Delta l)}{\Sigma \Delta G \cdot \sin \alpha}$$

results for the required anchor strength A (kN/m) the expression

$$A = \frac{\text{requ. } \eta \cdot \Sigma \Delta G \cdot \sin \alpha - \Sigma(\Delta G \cdot \cos \alpha \cdot \text{tg}\varphi' + c' \cdot \Delta l)}{\sin \alpha_A \cdot \text{tg}\varphi'_A + \text{requ. } \eta \cdot \cos \alpha_A}$$

whereby the required safety coefficient $\eta = 1.25$.

ΔG = weight of lamella
Δl = length of lamella within slip surface
α = gradient of lamella base to horizontal
α_A = angle between resultant A and chord of slip surface
φ' = friction angle in area A

The length of the adhesion zones of the anchors to be positioned beyond the slip surface was assumed to be 10 m.

In the design, all stages of construction work had to be taken into account: As long as the bore-pile panels were not anchored, they had to be considered as fixed at the bottom and free-standing. To distances of 5 to 10 m ahead of pile-boring operations, which were carried out progressing in one direction, a reduction of slope movements set in and only this made it possible to pull the casings.

For exploration purposes, the first pile was sunk deeper than called for by static calculations and shiny slickensides were found in the zone between 22 and 24 m. Aided by this finding and after a reconstruction of what was presumably the original valley bottom, slide circles a and b, shown in Fig. 6-68,

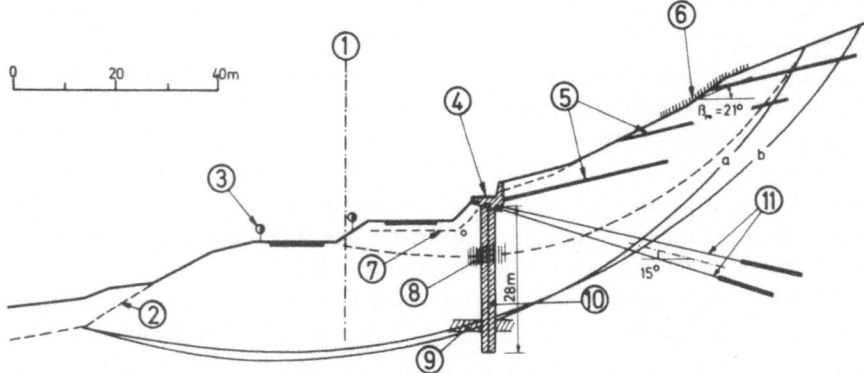

Figure 6-68 Section of slide area with stabilization measures; (1) axis of motorway; (2) assumed original valley bottom; (3) basis of geodetic measurements; (4) path on head beams; (5) drainage borings; (6) overgrowth; (7) filter fabric; (8) soft zones; (9) old slickensides; (10) bore-pile panels; (11) alluvial anchors.

were determined, yielding for a mean assumed friction angle $\varphi' = 18°$ and $c' = 0$, safety coefficient $\eta = 1$. Since a reactivation of slip surfaces could not be excluded, the design was changed at short notice; pile length was increased from 21 to 28 m and the reinforcement strengthened. The lowermost meters of each pile boring drilled without casing were inspected *in situ*.

Since the water flowing through the slope had a high share of carbonic acid, extra precautions had to be taken to protect the material of the piles. Blast furnace slag cement containing at least 45% slag, at least 350 kg/m^3 binding agent, and a concrete liquefying additive was used for the concrete. The resulting disadvantage was that this concrete has a low initial strength, entailing the danger that the green piles would shear off. Therefore the uphill piles, which were constructed first, were given a larger diameter. At the time the bore piles were constructed, the slope moved near the surface by 4 to 20 cm per week. After completion of the bore piles, the head beam was concreted in sections of 28 m, the retaining wall was erected, and the whole area planted to stop movements of the top soil. To achieve the desired safety, the retaining wall was anchored (Enderli, 1977) but alluvial anchors had to be used in spite of the occasional presence of rock. The time span between boring, placing the anchors, and grouting had to be kept as short as possible, since the talcy grauwacke-slate quickly softened in contact with water or formed lubricating layers. Anchor strength was increased by compacting during primary grouting and fracturing of the adhesion zones (24 hours after primary grouting); pressures in the range of 25 to 35 bar were used. Without these additional measures the required pretension strength would not have been achieved or the admissible creep deformation would have been exceeded. Furthermore, it was necessary to protect the concrete piles against the very agressive water coming from the slope.

Anchoring operations were supplemented by core and drainage borings and behavior of completed sections of the bore-pile wall was constantly monitored by stress–strain measurements. Each of the 28-m-long sections was provided with two measuring anchors and the anchor heads contained measuring plates.

6.2.9.4 Anchored Wall Constructed from Top to Bottom—Anchor Wall near Peggau, Styria, Austria

The alignment for the section Peggau–Frohnleiten of the four-lane freeway between Graz and Bruck required a cut through a very steep mountain ridge. Extensive constructions above ground were the only alternative to earth-moving operations and were rejected from the start for reasons of economy and because they would have marred the landscape.

The slope had an average gradient of 40° to 45°. The surficial layer consisted of loamy (sandy-clayey silt) talus and no outcropping bedrock was evident. To explore subsoil conditions, three shafts were sunk at the crown of the slope to depths of 3 m at distances of 15 m, and four core borings about 20 m deep were carried out. Contrary to expectations, no bedrock was found but there was talus cemented by silty-sandy components containing boulders and in some places sand and silty sand. The particles of the talus were all angular;

it was to be classified as loose soil and its angle of internal friction was in the range of 35° to 45°. Even if a certain cohesion was assumed in the calculation, the original slope had to be considered very steep and near limit equilibrium. The type of cut required for conventional retaining walls would have caused the greatest difficulties and it would have been virtually impossible to brace the over-15-m-high cut in the slope. Furthermore, a retaining wall without anchors would have to be designed with a profile of considerable dimensions.

From the soil on the site a free-standing height of 3 to 4 m could be expected. Therefore a construction method was chosen which was still relatively new in 1972 (it was first tried in 1968 during the construction of the Brenner Freeway; the method was to construct a retaining wall from top to bottom with horizontal strips of concrete slabs (Fig. 6-69). These reinforced-concrete slabs (6.0 × 2.5 m) were anchored with two or three anchors, depending upon local requirements. The cut was up to 17 m high, 206 m long, and its slopes had a 7:1 inclination (vertical to horizontal). The required length and strength of the anchors was determined after Schindler (1969), resulting in lengths between 11.0 and 21.5 m and a pretension strength between 380 and 880 kN per anchor, whereby anchor strength increased from top to bottom.

The question came up of how to achieve by the cheapest method the strongest possible interlocking between one strip of reinforced-concrete slabs and the next. This problem was satisfactorily solved by the shape and position of the

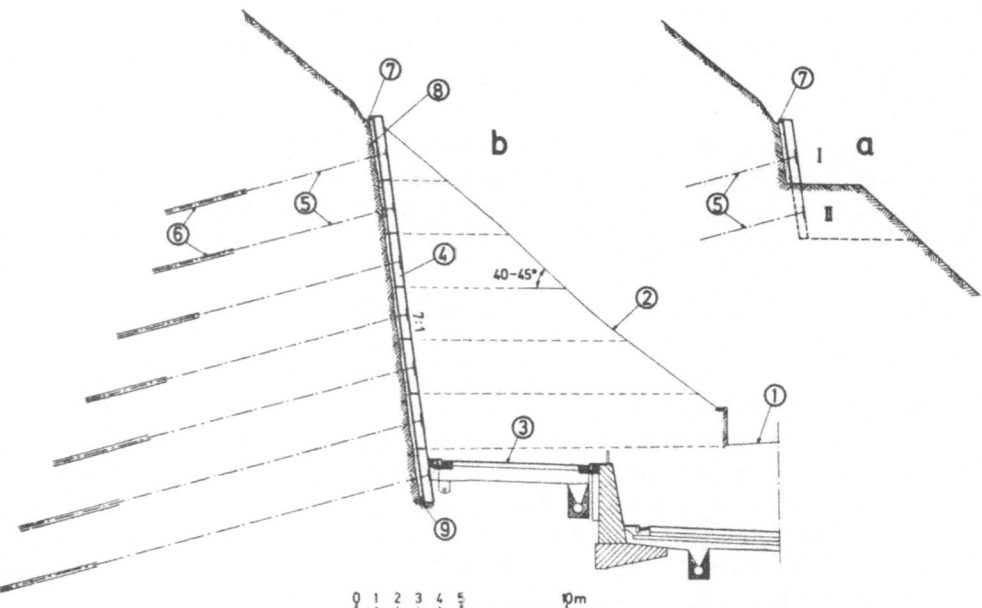

Figure 6-69 Section of anchored wall; a—stages of construction; b—completed section; (1) old motorway; (2) original embankment; (3) new road uphill of expressway; (4) reinforced-concrete slabs; (5) free lengths of anchors; (6) adhesion zones; (7) surface drainage by open channels; (8) single-grain concrete; (9) drainage pipes.

openings for pouring and vibrating the concrete. A layer of single-grain concrete was filled between concrete slabs and original soil to divert slope waters safely and quickly, to even out irregularities caused by cutting operations, and to protect the wall against capillary water. The surface waters were collected in an open drainage channel at the edge between anchor wall and natural slope.

Stages of Construction. Excavation for the individual strips was carried out as shown in Fig. 6-69a. Excavation for the first strip was carried out in one course over the whole length. Then the layer of single-grain concrete was applied, the anchors installed, the reinforced concrete slabs cast in place, and the anchors tensed.

Construction of the following strips II to VII proceeded as follows: First the soil below the already-anchored strip was excavated so that an earth wedge, with a gradient below that of the natural friction angle of the soil, remained standing to serve as temporary bracing. Then this earth wedge was removed over the length of one slab and this slab was constructed as described above. Next the span of one slab was jumped and the second slab constructed. After every other slab was in place, the remaining earth wedges were excavated and the rest of the slabs constructed. The anchor holes were bored, on principle, after the application of the filter layer, and the alluvial anchors were installed by the well-tried method of a specialized company, but it is not within the scope of this book to deal with such details.

After the reinforced-concrete slabs were in place and had set, the anchors were tensed in accordance with the respective testing regulations. From time to time geodetic control measurements are carried out at the anchor wall to monitor the stability of the construction. Where conditions are especially critical, measuring anchors are used to control anchor tensions periodically.

The main advantage offered by the above construction method is that natural conditions in the cut slope are only slightly disturbed. Any buildings situated above the cut remain untouched, something that is rarely possible when constructing conventional retaining walls because their excavation operations reach far into the surrounding area. Lastly, this construction method requires very little space at the foot of the wall. Of course it is economical only up to a certain anchor length.

6.2.9.5 Other Retaining-Wall Constructions

With "Krainer" crib walls, gabionades, reinforced earth, the new "Ebenseer" Wall, and soil nailing, a joint action of the strength of the fill (stone fill or compacted layers of soil) or the original soil and the tensile material (wood, wire-mesh, steel, etc.) is obtained. All of them permit an easy passage of any water contained in the slope and will also follow major settling differences without being damaged.

6.2.9.5 (I) "Krainer" Crib Walls. "Krainer" crib walls are constructed by alternately placing rows of round timber (diameter—25 to 30 cm) at distances of 40 to 50 cm parallel to the slope (runners—two lines, one at the face and the other at the inside end of the binders) and at a right angle to the slope (binders—placed at distances of 1.2 to 1.5 m). The binders should have an inclination to the slope that is slightly below horizontal; their lengths depend upon the designed thickness of the wall. The logs have grooves where they cross and are held together by steel clamps.

As one row of logs after the next is in place, the enclosed space is filled with soil from the site, which is then compacted, or with riprap (stone size in the range of 20 × 20 × 20 cm to 40 × 40 × 40 cm). The wall acts like a brick-work wall because of the combined effects of wood and soil or riprap, and will take up the resultant of its own load and the earth pressures, as well as additional forces from slide movements, and transmit them to the foundation joint.

This construction has the advantage that it is permeable and water pressures behind the wall are avoided. In addition, it will follow settling of a resilient subsoil without reduction of its stability.

"Krainer" crib walls are very economical as temporary constructions, provided wood is available in the vicinity, and are much in use for the construction of roads that serve farming and forestry purposes.

Under the general term *crib walls,* various companies offer walls similar to "Krainer" crib walls consisting of precast reinforced-concrete elements, which are now much in use where retaining measures are required, in particular in connection with the construction of roads. Lately some problems have occurred, but these were by no means due to the method as such. In some cases it was found that the concrete used for the precast elements was of poor quality; but for the most part failure of this type of wall was due to the fact that the second line of runners at the inside of the wall was omitted (Bhandari, 1980).

6.2.9.5 (II)   Gabionades. The basic idea behind the construction of this retaining body is more or less the same as with "Krainer" crib walls, i.e., loose, preferably round stones with diameters in the range of 10 to 20 cm are tied together by wide-meshed wire nets (width of mesh 4 × 4 cm, diameter of wire— approximately 3 mm) to form blocks (about 1 × 1 × 1 m). These blocks are piled up and act like a retaining wall. Just like the "Krainer" crib wall, this type of wall is permeable and will also adjust to large settling differences of the subsoil without being damaged. In areas where wood is scarce, it is very economical to use gabionades.

6.2.9.5 (III)   Reinforced Earth. This is a construction method for which standard regulations (codes of practice) already exist in several countries, e.g., France and the United States (Maluche, 1976).

Similar to the first two methods, this method is based on a combination of earth and tension-proof material, but in this case armoring tapes (lengths up to 25 m; width—40 to 120 mm; thickness—3 to 5 mm) are installed in the soil at vertical and horizontal distances of about 75 cm. These tapes take up the tension stresses and transmit them by friction to the fill; thus, soil and armoring tapes achieve a joint load-bearing capacity. On the exposed face either a permeable concrete lining (thickness—18 to 20 cm) or steel plates (thickness—3 mm) are fixed to the armoring tapes to prevent the backfill from flowing out. The flexibility of this lining allows the construction to follow large settling differences without damage.

In Washington in the United States, a road crossing a slope had been sliding down 9 m over a length of 61 m and was stabilized by two staggered retaining bodies of reinforced earth constructed one above the other. The upper retaining wall, which carries the roadway, is 6.70 m wide and 7 m high; the lower one, which stabilizes the slope, has the impressive dimensions of 16.7 by 12.0 and was carried out in sections to avoid further slides. The subsoil consists of silty clay and weathered siltstone on top of solid siltstone. To divert the water running down on the siltstone a filter layer was first spread. This construction was by far less expensive than the originally foreseen slope bridge with pile footings.

6.2.9.5 (IV)   New "Ebenseer" Wall. This is a new type of retaining wall which is especially well-suited for large dimensions. The exposed face consists of precast concrete elements with noncorrosive steel tapes fixed to their backsides. These tapes are laid in loops around the deflector elements, which are placed inside the soil of the original slope. As in the case of reinforced earth, this retaining wall is constructed in strips, whereby each strip is filled with soil from the construction site and compacted. The advantage of this wall versus reinforced earth is that the tensile stresses on the steel are not transmitted to the soil by friction but by way of the deflector body (similar to a dam built of circular cells), thus making better use of the steel tapes. The deflector body may be set back at random, thus achieving a better anchoring.

In 1977 model tests were carried out by my department at the Graz University that showed that the wall had a high load-bearing capacity and a great stress–strain reserve. The first practical experience was also gained in 1977 on a small construction site near Kremsbrücke, Carinthia, Austria. Further tests in practice are under way. In my opinion the "Ebenseer" wall is a development with a future, especially in view of its stability against earthquake action.

6.2.9.5 (V)   Soil Nailing. By this method, the soil is also provided with a steel reinforcement, i.e., a retaining body is created in which the soil itself takes over a load-bearing or retaining function; however, here it is not first excavated, then filled around the steel tapes, and mechanically compacted. It remains in place, and is provided with a reinforcement, e.g., threaded steel rods (diameter—22 mm), as excavation work progresses.

Stocker and Bauer (1976) report in detail about model and large-scale tests carried out to analyze load-bearing behavior and failure criteria of this composite system. This analysis covers different types of nails, nail lengths, and distances between nails; it also deals with design methods.

In a large-scale test a 7-m-wide and 6-m-high wall was constructed by the following method: First, a 1.5-m-deep cut was made and immediately lined with an 8-to-10-cm-thick reinforced shot-concrete skin. Then 3-to-4-m-long anchoring nails (threaded steel rods, St 42/60, diameter—22 mm) were rammed in at horizontal distances of 1.2 m with simultaneous grouting with cement mortar.

After the wall was completed to its full height of 6 m, the ground surface was loaded with 0.26 MN/m² at a distance of 3.0 m from the wall line. Under this load the top edge of the wall showed a horizontal dislocation by about 20 mm, the toe by less than 5 mm. Only after the lowermost row of nails was extracted and a 0.9-m-deep trench excavated below the wall did quickly progressing deformations set in under the same load.

In my opinion, this method may be used to stabilize embankments of construction pits as well as sliding zones in existing embankments, whereby soil nailing is not an anchoring in the conventional sense but rather an improvement of stability conditions in the soil.

6.3   Synoptic Description of Characteristic Landslides

As a summary to 6.1 and 6.2, Table 6-2 gives an overview of results of slide investigations in Japan and stabilization methods chosen. This table is of quite some help in the statistical registration and classification of landslides as well as the selection of stabilization measures; it deals with slides that occurred near the towns of Kanazava, Toyama, and Nagano (on May 8, 1947, the Zenkoji earthquake—magnitude 7.4 on the Mercali scale—caused about 44,000 landslides).

I visited the slide areas described in Table 6-2 and offer the following comments and classifications from the viewpoint of the degree of stabilization achieved:

(A)   Sliding slopes that have been virtually fully stabilized:
   (1)   Landslide on Road Nr. 8, between Toyama and Joetsy (not shown in Table 6-2)
   (2)   Kamihiramaru Landslide
(B)   Sliding slopes that were stabilized to a great extent but still show minor movements that are constantly monitored:
   (1)   Kamenose Landslide
   (2)   Kurumi Landslide
   (3)   Chausuyama Landslide
   (4)   Kodomari Landslide
(C)   Sliding slopes that have not yet been brought to rest, where soil movements at the surface and at different depths—mainly caused by fluctuations of the ground-water table—are measured and the effects of various stabilization measures registered:
   (1)   Sarukuyoji Landslide
(D)   Landslides that developed spontaneously in recent times:
   (1)   Ikadani Landslide.

The following common characteristics of the landslides are, in my opinion, to be noted:

(1)   With the exception of the Kamenose and Ikadani slides, the course of the slip surfaces runs over wide areas parallel to the ground surface,

meaning that they are predominantly translational slides (Skempton and Hutchison, 1969) and not rotational slides.

(2)   As a rule, the slip surfaces are at the zone of division between two different types of soil and are not governed by the strike and dip of the layers.

(3)   In all cases, water is present at the slip surface, mostly in very minor quantities—only like a film. The pore-water pressure of this water is to be seen as the main cause for the slides (e.g., the Chausuyama slide, where the water table was lowered 20 m by drainage and the mean annual movement reduced from 130 to 20 cm). The same applies to almost all other slides.

(4)   The development of cracks was a warning for almost all landslides. If such cracks appear, the potential slip surface must be located and stabilization measures initiated immediately.

(5)   Smaller and larger water pools were on the whole slide area but especially near its head. These pools continuously supply the water for the film at the slip surface.

(6)   Landslides mostly set in during or after heavy rainfall or during melting (in the area of the Ikadani slide snow reached a height of 6,780 mm in the winter of 1976–1977; on March 6 only 1,300 mm were left and by March 18 it was all gone). It is obvious that this sudden inflow of such large quantities of water strongly increased pore-water pressure.

(7)   Landslides frequently occur in areas where landslides have previously taken place. Such areas need to be kept under constant observation, even if they are at rest for the moment. Monitoring measures should include registering the pore-water pressure at the slip surface and, if possible, the electric potential differences between the layers above and below the slip surface, by periodic measurement from inspection shafts. In addition, the ground surface should be observed for any possible movements.

The stabilization measures to be carried out may be summarized as follows:

(1)   Since the pore-water pressure at the slip surface evidently causes the landslides, the water at the slip surface must be eliminated by all means and a further inflow must be prevented. This is to be achieved by:
       (a)   Drying out the ponds on the surface of the slide area, especially those at the head of the slide
       (b)   Closing the cracks and fissures with a mixture of sand, cement, bentonite, and water to prevent the penetration of surface waters
       (c)   Careful diversion of surface waters (rain and melting snow)
       (d)   Drainage by combined vertical and horizontal shafts and ducts, or drainage borings
       (e)   Elimination of electric potential differences at the slip surface by installing short-circuit conductors, after Veder.

Table 6-2

		KAMENOSE	KURUMI	KODOMARI
LOCA-TION	LONGITUDE (E)	135°40'20" – 135°41'5"	136°55'40" – 136°56'30"	138°0'5" – 138°0'15"
	LATITUDE (N)	34°34'40" – 34°35'10"	36°55'10" – 36°56'15"	37°6'10" – 37°6'25"
	PREFECTURE	OSAKA	TOYAMA	NIIGATA
MORPHOMETRI-CAL VALUE	LENGTH L_R/L_T (m)*	1,130/1,130 1,030/1,030 (max.)	1,500?/1,550 450?/480	250/400 140/140
	WIDTH W_R/W_T (m)			
	HEIGHT H_R/H_T (m)	216/216	265?/270	65/100
	GRADIENT G_R/G_T (°)	10.8°/10.8°	10.0°/9.9°	14.6°/14.0
	DEPTH D_{AV}/D_M(m)	40/75	25?/50	4.3/15?
	AREA A_R/A_T (ha)	54/55	68?/74	3.5/5.6
	VOLUME V (m³)	22,000,000	18,500,000	150,000
HISTORY OF MAIN MOVEMENT		Before 37,800 B.P. 1931 in Toge 1953 in Shimizudani 1967 in Shimizudani and Toge	? 1917 1926 1941 July, 1964	? · · March, 1963
TYPE OF LAND-SLIDE	(HATANO and OYAGI, 1977)	Bed rock and unconsoli-dated material slide	Bed rock and unconsoli-dated material slide	Unconsolidated material slide and earth flow
	(VARNES, 1975)	Rock and debris slide	Rock and debris slide	Earth slide and flow
RELATION BETWEEN DIRECTION OF MOVE-MENT AND GEOLOGIC STRUCTURE		*Nagare-ban* † Dip-ward	*Uke-ban* † Reverse to dip-ward	*Uke-ban* † Reverse to dip-ward
GEOLOGY	LITHOLOGY	Andesitic lava, tuff and tuff breccia; sandstone and conglomerate.	Mudstone and tuff	Mudstone
	AGE	Miocene (Tertiary)	Miocene	Miocene
PROPERTIES OF MATERIAL	MOVING MASS	Debris: silt, sand, blocks. rock masses	Debris: silt > clay, sand gravel, blocks. rock masses	Debris, earth: clay, silt, sand, gravel, ...
	NEAR SLIP SURFACES	Clay, sandstone, tuff ..... Yield point C: 0.3 – 0.9 kg/cm² ϕ: 19 – 22°	Tuff in natural water content C: 0.23 – 0.78 kg/cm²	
	UNDER SLIP SURFACES	Rock Reverse calculation C: 0.2 kg/cm² ϕ: 11°	Tuff in 10% larger than natural water content C: 0.04 – 0.15 kg/cm²	
RATE OF MOVEMENT	RECENT (AVERAGE)	~ 5 cm/year	1 – 7 cm/year	?
	MAXIMUM	0.5 m/day, in 1931	2 – 3 m/sec. in 1964	0.03 m/sec, in 1963
CONTROL WORKS		Surface drainage Drainage tunnels Drainage wells Steel tube piling Deep foundation Soil removal	Surface drainage Drainage wells Composite borings Steel tube piling Sabo dams Hillside removal	Drainage Steel tube piles Retaining wall works
LAND USE		Farmland	Farmland, forest	Farmland
PROBLEMS TO SOLVE		1) Relation between two areas, Shimizudani and Toge. 2) Influence of design dis-charge of the R. Yamato to the stability of Toge area.	1) Future instability in the area behind main and lateral scarps.	1) Future instability in the area behind scarps.

Note: *: L_R: horizontal length of root area; L_T: horizontal length of landslide (whole area); W_R: width of root area; W_T: width of landslide; H_R: height of root area; H_T: height of landslide; G_R: gradient of root area (crown to foot); G_T: gradient of landslide (crown to toe); D_{AV}: average depth of landslide; D_M: the maximum depth of landslide; A_R: area of root area; A_T: area of landslide.

†: The words used by mining engineers in Japan.

††: New words adopted in this guide-book.

SARUKUYOJI	KAMIHIRAMARU	CHAUSUYAMA	IKADANI
138°19'30" − 138°20'30" 37°0'10" − 37°0'45" NIIGATA	138°19'25" − 138°19'40" 36°56'5" − 36°56'35" NIIGATA	138°6'35" − 138°7'48" 36°35'2" − 36°35'30" NAGANO	136°58'0" − 136°58'58" 36°54'48" − 36°55'10" TOYAMA
180/1,700 200/250 60/280 18.4°/9.4° 5°/? 3.6/43 2,150,000	480/950 300/400 140/240 16.3°/14.2° 5/10 14/30 1,500,000	800/1,900 200/450 (max.) 120/300 8.5°/9.0° 20/80 16/38 9,000,000	1,230/1,250 250−400/300−400 232/232 10.7°/10.5° 30/40 37/38 11,400,000
? . 1,100 . 1,900 . . .	? Spring, 1932 Spring, 1945 Spring, 1958 Dec., 1963 Dec., 1964 Apr., 1969 Apr., 1970 Apr., 1974	May 8th, 1947: Zenkoji Earthquake 1884: Cracks 1898: Uplift at foot 1911~: Remarkable movement 1930~: Flow	1907 Large-scale slide 1935 ditto 1938 ditto 1939 ditto 1954 ditto 1957 Small-scale slide 1961 ditto 10:45 a.m. on March 29, 1977.
Unconsolidated material Creep Earth slide (or Creep?)	Unconsolidated material slide and earth flow Earth slide and flow	Unconsolidated material slide and earth flow Earth slide and flow	Bed rock and unconsolidated slide. Rock and debris slide
Naname-ban †† Oblique to dip-ward	Yoko-ban †† Parallel to strike (Normal to dip-ward)	Yoko-ban †† Nearly parallel to strike (Normal to dip-ward)	Nagare-ban † Nearly dip-ward.
Mudstone	Mudstone	Mudstone, sandstone and tuff	Mudstone, sandstone and tuff
Miocene	Miocene	Miocene	Miocene
Earth: silt=clay > sand W: 30 − 40% C: 0.11 − 0.28 kg/cm² φ: 19° − 31°	Earth: silt > clay > sand W: 26 − 31% C: 0.12 − 0.29 kg/cm² φ: 2° − 9° qu = 0.43 − 0.59 kg/cm²	Debris; sand > clay, gravel, blocks. Rock masses.	Debris. Rock masses.
Clay ≧ silt > sand W: 40 − 50% C: 0.23 φ: 12°	Silt > clay > sand W: 30 − 32% C: 0.15 − 0.19 kg/cm² φ: 9° − 12°		
Weathered mudstone W: ~ 30% C: 0.42 − 0.49 kg/cm² φ: 38° − 46°	Weathered mudstone W: 14 − 19% qu: 13 − 32 kg/cm²		
1 − 3 m/year	Creep-slide: 1 − 2 m/year Flow: 2 m/sec, in 1970	0.8 m/year 93 m/year, 1933 to 1935	0.5 m/sec ?
Surface drainage Underground drainage Veder's work	Surface drainage Drainage wells Sabo dam Check dams Steel tube dams	Surface drainage Deep wells Drainage wells Drainage tunnels Curtain wall Steel tube piles Sabo dams	Surface drainage Horizontal borings
Farmland	(Farmland), Badland	Natural botanical and zoological garden	Farmland
1) Will the weathered bedrock under the slip surface be instable or not? 2) Further rapid slide behind the main scarp by earthquake, heavy rainfall etc.	1) Was the domain I an ancient large-scale slide mass? 2) Dynamic relation between the domains I and II.	1) Bedrock structure of the landslide area.	1) Further instability of the slide mass. 2) Future instability of unmoved mass in the lower portion of the landslide.

Reference: S. Hatano and N. Oyagi (1977): Methods of survey for landslides and slopes. (in Japanese). Chiri (Geography), Vol. 22, No.5, pp.56−71.
D.J. Varnes (1975): Slope Movements in the Western United States. Mass Wasting 4th Guelph Symposium on Geomorphology, 1975, edited by E. Yatsu, A.J. Ward and F. Adams. pp.1−17.

(2)   In addition, movements of the slide body need to be monitored by surveys of surface measuring points, by extensometers on the ground
surface, by inclinometers, by stress–strain measurements at the slip
surface, etc.

(3)   From the results of these measurements conclusions may be drawn as
to the stability of the slide or any further stabilization measures that
are required.

6.4   Additional Stabilization Measures

Protective constructions against snow and debris avalanches are additional
measures that are only required under special circumstances. These constructions, which often involve a major effort, belong to a specialized field and are
not within the scope of this book.

But in any case the seeding and/or planting of the slide area is an important
part of all stabilization measures discussed in this chapter and should always
be carried out after the completion of construction work (Gray, 1970; Rice *et
al.*, 1969).

Vegetation may significantly increase the stability of a slope susceptible to
slides or a slope that has already been stabilized by other methods. Vegetation
strengthens the surficial layer by

(1)   Reducing the amount of water running down the slope surface
(2)   Drying out the layer to some degree
(3)   Strengthening the layer by the lattice of roots

It is a fact that soils carrying grass have about 10% lower water content to
depths of about 2 m than soils that do not; where bushes grow this effect even
reaches to a depth of 3.5 m. On the other hand, sod will keep the soil from
drying out and thereby prevent the development of shrinkage cracks.

Regarding the stabilizing action of bushes and trees, those trees are most
effective that withdraw the greatest amount of water from the soil, e.g., birches,
alders, aspen, poplars and willows; less suitable are coniferous trees. However,
care has to be taken that trees do not overburden the head of a slope as they
grow older and heavier; therefore, those with a trunk diameter of more than 30
cm should be felled and replaced by younger trees.

A complete deforestation must be absolutely avoided because it changes
ground-water conditions; it also robs the soil of the stabilizing effect of the
roots, and slides may develop again. Roots will grow the strongest where they
find the highest humidity; by this natural process roots will seek the critical
zones, i.e., the water-rich zone between brown and blue clay, and there withdraw the water and contribute further to the stabilization. If trees are felled
and no new trees are planted, the water content at the critical zone of division
increases and within a short time, slide phenomena will set in. The author has
stabilized many slides of this origin by ramming in steel rods as short-circuit
conductors.

Where good vegetation is lacking, erosion grooves develop in the surface which aid the undesirable penetration of water into the soil. Here the damage done to the sod by grazing cattle should be pointed out. It is especially cows that rip out whole bunches of grass together with the roots, and by their weight damage the sod and open drainage trenches. Sheep, on the other hand, feed on the young shoots of trees and bushes, which makes it very difficult, if not impossible, for the forester to maintain and rejuvenate his forest. It is very hard to bring back to life a formerly green mountainside that has turned into a karst.

An excellent process to achieve a quick growth of grass is the Schiechtel method. A mixture of bitumen, straw and grass seeds is sprayed on the ground surface. The bitumen makes the straw stick to the ground and forms with it a firmly attached nutritive substratum for the quickly growing grass. Wicker-work of willow rods placed in a chessboard pattern or a grid of concrete bars provide an excellent additional hold. Larger cracks and fissures at the surface should be closed and if possible sealed watertight with a mixture of cement, sand, bentonite, and water (see 6.1.4.1 (I)).

To avoid damage from rockfall, steep slopes should be provided with retain-ing nets of wide-meshed nylon, so that vegetation will fully cover the area.

6.5 Durability of Stabilization Measures

From the stabilization methods described in this chapter, a permanent effec-tiveness may be expected, provided they are properly designed and carefully carried out. Their life span corresponds absolutely to that of other constructions.

As far as my experience goes, stabilization measures will mainly fail where drainage was designed and carried out on the basis of inadequate investigations which led to wrong positions and/or depths of the individual drainage lines. Another cause for ineffectiveness is that instead of the modern technical means available today, outmoded material is used, e.g., concrete and brick ducts are still used in some places. These pipes do not have a filtering effect and therefore do not guarantee a lasting effectiveness, even though their use may be more economical.

Where very extensive and expensive stabilization projects are involved, it may serve the purpose not to start out with the full design, provided constant controls are instituted. If it turns out in the course of construction operations that all of the designed measures are required to achieve the desired stabili-zation, the remaining measures still can be carried out afterwards. This is the repeatedly discussed "observation method" recommended by Terzaghi.

6.6 Symbols Used in Soil Profiles

A number of illustrations in this chapter show soil profiles as they were obtained by exploratory borings in and around the slide areas. The symbols used alongside the columns of the soil profiles are those contained in DIN 4023 (German industrial standards). For ready reference they are listed below, together with the types of soil that they stand for:

		Grain Sizes
G	gravel	$>$ 2.0–63.0 mm
sG	coarse gravel	$>$ 20.0–63.0 mm
mG	medium gravel	$>$ 6.3–20.0 mm
fG	fine gravel	$>$ 2.0–6.3 mm
S	sand	$>$ 0.06–2.0 mm
gS	coarse sand	$>$ 0.6–2.0 mm
mS	medium sand	$>$ 0.2–0.6 mm
fS	fine sand	$>$ 0.06–0.2 mm
U	silt	$>$ 0.002–0.06 mm
T	clay	$\leqq$ 0.002 mm
Mu	vegetable earth	
o	organic admixtures	
A	fill	
X	stones	
Y	boulders	
Z	rock	

▽—Position of ground-water table at time of boring
▼—Position of ground-water table after stabilization
☐—Points of extraction of soil samples
⊠—Points of extraction of soil samples for field and laboratory tests

Capital letters	—Main component with a share of 40 to 60% by weight; if two types of soil contribute shares in that range, capital letters are shown for both soils and connected by a hyphen (e.g., S-G for sand and gravel).
Lowercase letters	—Secondary components; placed after the capital letter of the main component (e.g., U,fs,t for finely sanded, clayey silt)
Lowercase letters with bar	—Major secondary component with a share of 30 to 40% by weight (e.g., $T,\bar{u}$ for clay containing between 30 and 40% silt)
Lowercase letters with apostrophe	—Minor secondary component with a share of less than 15% (e.g., G,s' for gravel with some sand)

7 Physical Chemistry of Landslides in Silt and Clay Soils
by F. Hilbert

7.1 Introduction

Basically, there are two different categories of landslides. To the first category belong the slides with purely mechanical causes (e.g., an increase of hydrostatic pressure, erosion, or freezing) without change in the chemical properties of the soil. To the second category belong the slides that are not caused by a change of mechanical forces but by a change in the physical or chemical properties of the soil. Perhaps the most dangerous type of slides occurs because the soil loses its original strength in part (slip surfaces) or as a whole. In extreme cases even a liquefaction may set in and previously solid soils will behave like fluids.

Soils susceptible to this type of slide do not consist of macrocrystalline particles but (at least in part) of colloidal particles (range of size between 0.5 μm to 1 nm). The cause and the phenomenology of such slides can only be explained by taking into account the crystallographic and the chemical structure of the colloidal particles and the special properties caused by their small size.

In almost all cases, a gradual or sudden decrease of shear strength in a soil is due to the content of clay minerals. These are responsible for the effects and the mechanism of water absorption and desorption, ion exchange, swelling, plastification and thixotropy. In almost all of these phenomena electrokinetic and reduction/oxidation mechanisms play a significant part. Therefore, the properties of colloidal systems in general and of clay minerals in particular will be presented in the following sections.

7.2 Highly Disperse (Colloidal) Soil Components

7.2.1 Colloids—Common Properties

The term *colloids* was originally used for silicate matter. Already a century ago it was recognized that hydrated silicic acid behaves like animal glue under certain conditions. On account of this resemblance, *colloid*, the Greek word for glue, was chosen as the name for a whole group of substances with more or less similar properties.

The basic characteristic of colloids is the minimal particle size. Every substance can exist coarsely disperse (particle size > 0.5 μm) as well as colloidally disperse (0.5 μm $>$ particle size > 1 nm). In a colloid not only the properties of the bulk substance are determining the behavior; more important are the special properties of the surface. Considering the ratio of volume to surface it is easy to realize that the properties of the surface become more and more important as the particle size gets smaller and smaller. A cube with an edge length of 1 cm has a volume of 1 cm^3 and a surface of 6 cm^2. When this one cube is split into tiny cubes with an edge length of 100 μm, the total volume will still be 1 cm^3, but the total surface of all cubes is already 0.06 m^2. Reducing the edge length to the colloidal range of 10 nm, one gets 10^{18} cubes, still with a total volume of 1 cm^3, but a surface of 600 m^2.

The above makes it quite plausible that the "specific" surface determines whether a substance is classified as colloidal or not. The specific surface is by definition the real surface area per unit volume (as demonstrated by the above example of the split cube) or the real surface area per unit weight (which is proportional to the surface area per unit volume). In practice, the specific surface is usually stated in $m^2 \cdot g^{-1}$, because for several reasons, which are of no significance here, it is rather difficult to determine the specific density of colloidal substances.

One of the most important consequences of this classification by specific surface is that every substance must be considered as a colloid which is of colloidal size in any of the dimensions of space. Therefore the filamentary molecules of synthetics or of cellulose have to be considered as being colloidal, even though their lengths are far outside the colloidal range. What is much.more important in this context, the plateletlike particles of clay minerals belong to the class of colloids because their thickness is in the colloidal range, though their length and width are much larger. The photographs taken by a scanning electron microscope (Figs. 7-1 and 7-2; from Wagner, 1974) convey an impression of the size and shape of such particles. Colloidal substances may be amorphous (without defined crystal lattice) as well as crystalline. Almost all colloidal clay minerals are crystalline (the platelet crystals are very clearly visible in Figs. 7-1 and 7-2), while synthetic colloidal silicates are mostly amorphous.

A highly disperse substance need not be a solid; there are gaseous and liquid colloids too. From the viewpoint of our discussion, of course, the solid highly disperse substances in natural soils are the most important.

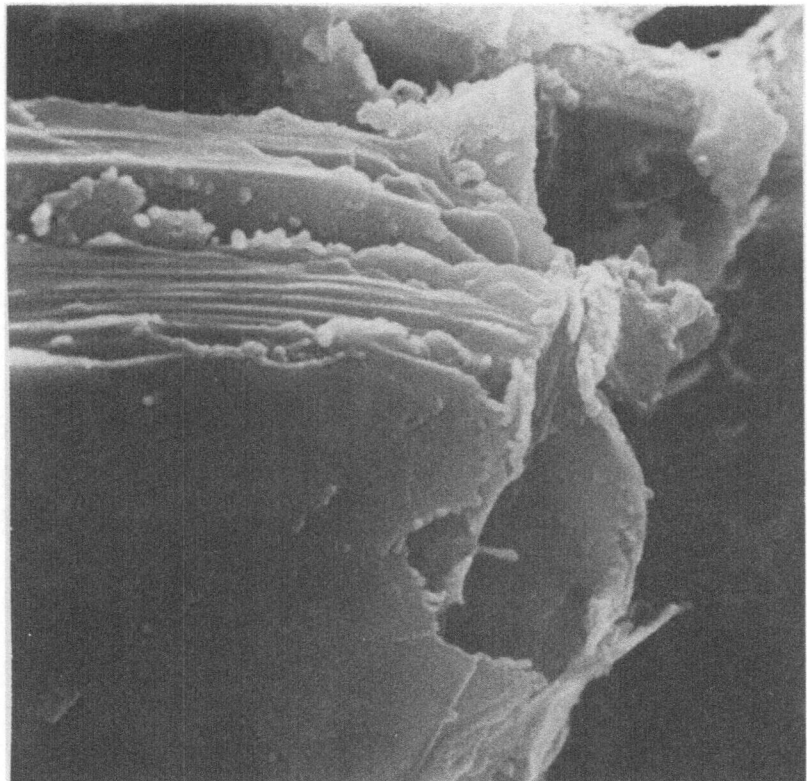

Figure 7-1   Platelet crystals, aligned face-to-face, in a sample of undisturbed brown, stiff clay; low water content, high shear strength. Enlargement by electron microscope 1:14,000.

The term *highly disperse substance* generally refers to a substance consisting of very small particles with a very large specific surface. In the normal case these particles are finely dispersed in a second, homogeneous phase. If there is an excess of the homogeneous phase, the whole system is called a colloidal solution, a colloidal dispersion, or a *sol*.

Coagulation (flocculation, flocking out) may turn such a sol into a *gel*; the colloidal particles settle together forming macroscopic aggregates which begin to sink as they become heavier than the homogeneous phase.

In principle, colloids may be formed in two ways. Colloidal dimensions may be reached by the bonding together of smaller particles (molecules, ions) thereby forming larger aggregates (colloidal molecules, colloidal crystals, colloidal amorphous solids, colloidal droplets of liquid and colloidal gas bubbles). Most of the natural silicates were generated in this way. On the other hand, colloids may be formed by the splitting up of larger particles. A very considerable amount of energy is required for mechanical disintegration to colloidal size and in practice one gets only very rarely a complete transformation of the

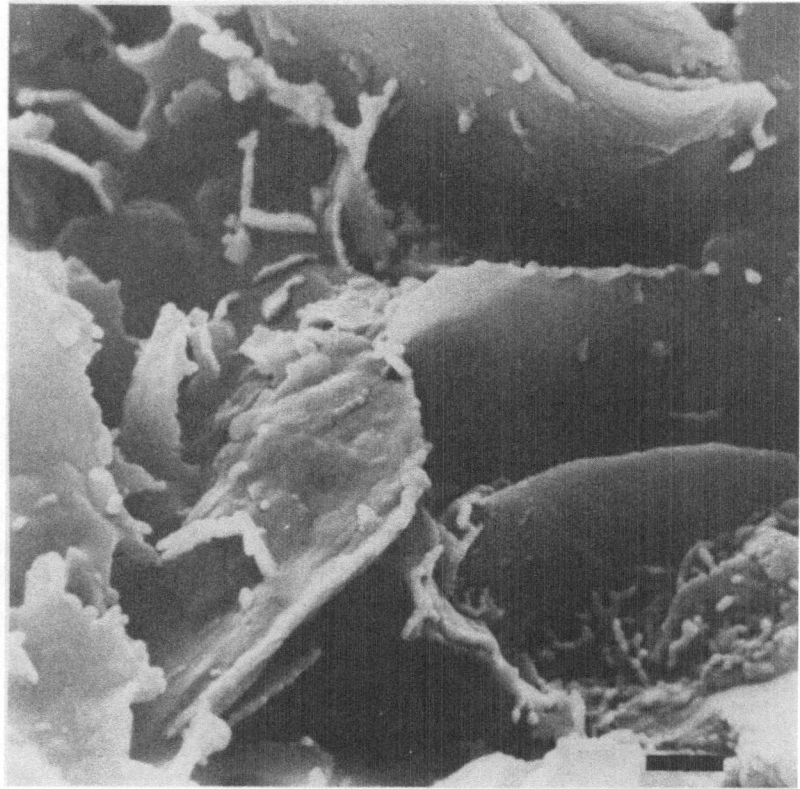

Figure 7-2 Unaligned platelet clay-mineral crystals from a slip surface (zone of division between blue and brown sandy-clayey silt). The crystals only touch edge-to-face, forming a house-of-cards structure; low shear strength, high water content. Enlargement by electron microscope 1:14,000.

original material into a colloid. However, this process is of great significance in industry (e.g., the manufacture of cement).

The properties of colloidal solids are very different from the properties of the same substances with macroscopic particle size. In water, particles with a diameter of 0.0001 cm and a unit weight of 3.5 are sinking at a rate of 4 cm per 24 hours, while particles with 0.0001 cm diameter sink only a few millimeters per day. If the particle size decreases further, soon the sinking rate cannot be measured any more. This difference may be used to separate coarser solids from colloidal solids by sedimentation, and also to separate colloidals into fractions by size or unit weight of the colloidal particles. Determination of colloidal particle size is also possible. Fractioning and particle size determination are carried out in high-speed centrifuges.

Dialysis is another method of separating dissolved matter from colloidal particles. This method also uses the difference in size, but here the colloidal particles are by far the larger ones. The much smaller dissolved molecules and ions (range of size 10^{-18} to 10^{-7} cm) are able to penetrate semiporous membranes made of parchment, collodium, or synthetics, while the larger colloidal particles cannot pass through the membrane at all or only very much slower. In a dialyzer (Fig. 7-3) containing in space (1) a colloidal solution and in the outer space (5) constantly renewed clear water, the dissolved small molecules and ions will diffuse out through the semiporous membrane, and after a sufficient time only the colloids and water will remain in (1). This dialysis principle is also of great significance for natural colloidal soils; here the soil itself is the semiporous membrane, i.e., the soil layer adjacent to ground water or surface water, being itself a colloidal disperse gel, only permits the passage of dissolved molecules or ions. In this way, dissolved salts (and also adsorbed ions—see below) contained in the pore water may be washed out of the soil after sufficient time. This process may significantly change the properties of the soil while it remains a compact solid. Therefore, dissolved substances may be removed from the soil without dispersing it in water. The dialysis process is reversible: if space (5) of Fig. 7-2 is filled with a real solution, e.g., one of ordinary salt, the sodium and chloride ions of this solution will diffuse into space (1) and a colloid present there will take up salt. This uptake of dissolved salts is of great importance for natural soils. Just as the removal of dissolved salts, the uptake of salts, acids, or bases changes the properties of the soil.

Both phenomena, dialysis as well as its reversal, are strongly influenced by electric fields if the particles migrating out of or into the colloid carry electric charges, i.e., are ions. Especially in the context of this discussion, the classification of *colloids* as *hydrophilic* or *hydrophobic* is of importance. Hydrophilic colloids have a low surface energy against water and therefore are easily wetted. Substances that are related to water are hydrophilic and therefore are especially capable of bonding water molecules more or less firmly at their surface. This creates a water sheath adhering to the surface, which remains intact even when the particles coagulate from a colloidal solution (conversion from sol to gel), so that in an apparently dry and solid gel the particles may not be in direct contact but separated by a water film. Clearly, the ions generated by the dissociation of water $H_2O \rightarrow H^+ + OH^-$, the positively charged hydrogen ions H^+ and the negatively charged hydroxyl ions OH^-, have a very strong

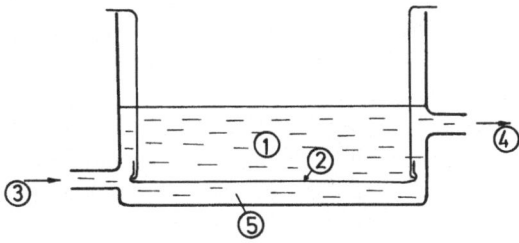

Figure 7-3 Dialyzer for ion exchange in colloids; (1) inner container with colloids; (2) semiporous membrane (permeable for ions, impermeable for colloidal particles); (3) inflow of outside solution; (4) outflow of outside solution; (5) container for outside solution.

influence on hydrophilic colloids. The conversion from sol to gel, and the reverse process, depend upon the concentrations of hydrogen ions and hydroxyl ions. Consequently, the pH-value of the surrounding water is of great significance for all hydrophilic colloids. In chemistry the pH-value is used as a measurement of the concentration of H^+ or OH^- ions and approximately corresponds to the negative logarithm of the H^+-concentration (pH $\cong$ $-\log c_{H+}$; see 7.2.2.3.2). Since virtually all colloidal soil substances are hydrophilic, the properties of hydrophilic colloids are of special importance for our subject. Hydrophobic colloids hardly ever occur in nature because they are very unstable and their particles tend to combine again.

7.2.2 Clay Minerals and Their Properties

All clay minerals are secondary formations, i.e., they were formed from the weathering products of mechanically or chemically decomposed primary rocks (magma). The chemical decomposition of primary rocks is caused by water containing carbon dioxide and oxygen. Other acids of natural origin (e.g., humic acid from the decomposition of organic substances) may play a part in the process. The decomposition products of alumina silicate and magnesium silicate are the most important ones in our context. Weathering of these silicates usually leads to complete destruction of the original crystalline structure. Formation of new minerals often takes place on the site of decomposition from silicic acid, aluminum hydroxide, and other weathering products present in colloidal solution. For example, kaolinite is formed by a chemical transformation from the primary mineral potassium feldspar, if the silicic acid is partially leached out and the potassium completely leached out. In other instances only some of the constituents of primary rocks are dissolved and carried off by water, while the rest remain in place as so-called residual material, or are carried off mechanically.

Mechanical transportation by water often causes a distinct separation according to particle size and specific weight, because naturally the smallest and lightest particles are carried the farthest. The components leached out from primary rocks may precipitate again on another site, either due to chemical reactions or because the water evaporates or cools off, forming the so-called separation sediments. The kaolinite mentioned above stands between residual and separation sediments because the chemical reaction leading to its formation takes place on the site of decomposition.

When solid material is transported mechanically by water, depositions at first form loose sediments, such as rubble, gravels, sands, and silts. When these loose deposits are covered by further sedimentation or material from eruptions, then consolidation sets in caused by increasing pressure and temperature; these processes are called *diagenesis*. For example, when clay sludge is dehydrated by pressure, the mostly plateletlike particles align themselves at right angles to the direction of pressure. Because the water is partially removed and, owing to the face-to-face alignment, the bonding forces between the particles increase and the sludge is consolidated. This consolidation, however, will be reversed when the consolidation pressure is removed and water is again taken up. Besides this consolidation, the formation of new minerals takes place (e.g., mica). When pressures and temperatures are higher yet, as may be the case with overthrusts and overfoldings in the course of mountain formation, a metamorphosis of the sediments sets in; a typical product is slate rock. Diagenetic changes may also be effected by uptake of colloids or salt solutions, e.g., the cementation of sandy sediments by precipitation of silicic acid.

Clay minerals mostly belong to the so-called layered silicates; other minerals in clay are minor constituents or inactive minerals like quartz or limestone. Regarding deposits, one distinguishes between kaolines (main components—kaolinite, halloysite; formed at the site—see above), bentonites (clays with a high content of montmorillonite—sedimented from water), and clays proper, which are also deposited as very finely-grained sediments (main components—kaoline minerals, montmorillonite, illite, quartz). Mixtures of clay and limestone may be found with proportions ranging from pure clay to clay marl, marl, and lime marl to almost pure limestone; the shear strength and stability against sliding increases with the increasing lime content (Broms and Boman, 1976, and 6.1.5.3). Table 7-1 lists the most important clay minerals.

Obviously, the clay minerals that increase their volume by uptake of water (swelling) and simultaneously decrease in shear strength, are the ones mainly responsible for landslides. As shown in Table 7-1, minerals of the pyrophillite-talcum group, micas, and brittle micas show no or very little swelling when in contact with water. In connection with landslides, these minerals are only of importance because their cohesion and shear strength values are rather low on account of the platelet particle shape and low bonding forces between particles.

The minerals of the other groups are the ones that are decisive for landslides, and here in particular the montomorillonites, beidellites and illites. After sedimentation these minerals may be dehydrated, compacted, and consolidated by heavy pressure from superimposed soil or rock layers. When the load is removed, water can be taken up again. This water gets bonded to the colloid particles, in part physically and in part chemically. Water penetrates between the platelet particles, causing a large increase of volume and a large reduction of the bonding forces between particles. Dry montmorillonite already absorbs so much water from water-saturated air that its volume increases by 30% and it becomes plastic; in the process, 1 g of montmorillonite absorbs about 1 g of water. When in contact with liquid water, the water content increases to 5 g of water for every gram of montmorillonite, the volume increases up to 20-fold and shear strength becomes practically zero; the result is a substance that flows almost like water. Why and how this change takes place will be demonstrated in the following, using montmorillonite as example.

7.2.2.1 Montmorillonite as Example for Layered Silicates Capable of Swelling

Montmorillonite consists of two constituents which are held together by relatively weak electrostatic attractive forces: the actual silicate crystals with the

Table 7-1 Main clay minerals

Group	Name	Swelling with Water	Ion Exchange
Kaolinite-Antigorite	kaolinite dickite nacrite	moderate	moderate
	antigorite chrysotile	moderate	moderate
	halloysite	medium	medium
Pyrophyllite-Talcum	pyrophyllite talcum minnesotaite	none	none
Micaceous: a) Montmorillonites	montmorillonite volkhonskaite hectorite	very strong	very strong
b) Beidellites	beidellite nontronite saponite sauconite pimelite medmontite	very strong	very strong
c) Vermiculites	vermiculite jefferisite	strong	very strong
d) Illites	illite glauconite	medium	moderate
e) Micas	muskovite paragonite phlogopite biotite	moderate	small
f) Brittle micas	margarite ephesite xantophyllite	very small	very small
Chlorite-Sudoite	derived from phyrophyllite or talcum (e.g. talcum chlorite)	medium	medium

chemical formula $[Al_4Mg(OH)_6(Si_4O_{10})_3]^{-1}$, carrying a negative charge (symbolized by an elevated -1), and the positive sodium ions Na^+ contained in the water layer between the platelet silicate crystals. Figure 7-5 shows a section through a number of such platelets and the interposed water layers. The aluminum magnesium silicate layers carry a negative electric charge and would repulse each other without the positive sodium ions. The sodium ions just compensate the negative charge of the silicate layers, so that the whole pack of layers is electrically neutral. The silicate layers do not touch directly; what weak cohesion they have comes from the intermediate film of water molecules, the forces of adsorption, and the hydrogen bridge bonding. As illustrated in Figs. 7-4 and 7-5, respectively, the negative charges of the silicate layers are localized in certain spots and therefore it is understandable that the position of the layers in relation to each other is of importance for the cohesive strength of a pack of layers. It is also evident that higher-valency ions, such as bivalent calcium ions (Ca^{2+}), trivalent iron (Fe^{3+}) or aluminum ions (Al^{3+}), must cause a much stronger cohesion between the silicate layers and thereby a much greater strength. This is illustrated in Fig. 7-5.

7.2.2.2 Mechanism of Water Uptake and Swelling in Clay Minerals
A stronger cohesion between the platelets, as caused by bivalent cations, apparently counteracts the penetration of further water molecules: the water that enters is taken up between the layers; consequently, a stronger cohesion between the layers reduces the ability for water uptake. Contrary to that,

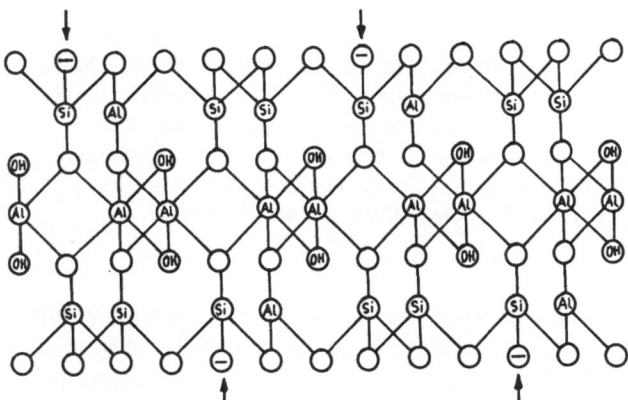

Figure 7-4 Diagram of structure of a layered silicate. The blank circles stand for oxygen ions. This section through a layer must be imagined to extend vertically to the plane of drawing. At the points marked by arrows are negative charges because Si (four positive charges) is replaced by Al (only three positive charges). The structure of the diagram corresponds to muscovite $[Al_3(OH)_2(Si_3O_{10})]^- \cdot K^+$ (potassium ions compensate for the negative charges). If one Ca^{2+} takes the place of two K^+, one obtains margarite (brittle mica). Montmorillonite has basically the same structure, but still a (significantly lower) replacement of Si by Mg^{2+} (instead of Al^+) takes place according to the formula $[Al_5Mg(OH)_6(Si_{12}O_{30})]^- \cdot Na^+$.

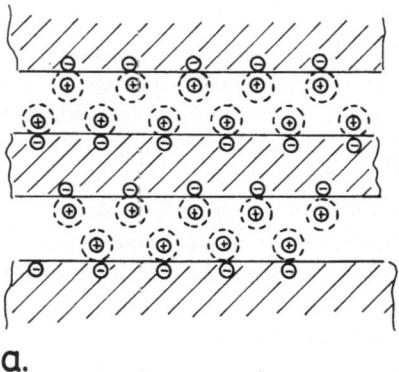

a.

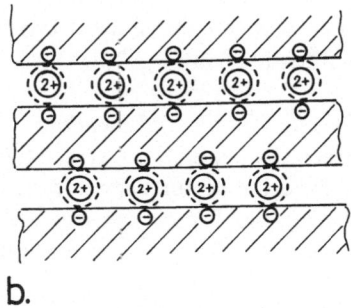

b.

Figure 7-5 Compensation of electric charges in layered silicates. a—Compensation of charges by monovalent cations: low cohesion between layers on account of the repelling effect of the positive charges of the cations and the larger distance between layers, caused by the stronger hydration of the monovalent cations. The sheath of water molecules around each cation is symbolized by dashed circles. b—Compensation of charges by bivalent cations: strong cohesion between layers, since cations are always bonded to two layers, holding them together, and a lesser distance between layers as the hydration of these cations is generally weaker.

monovalent cations between the layers will favor the water uptake because, first, cohesion between layers is weak, and, second, the positive ions and the negative electric charges of the silicate layers tend to attract water molecules. This tendency is due to the electrostatic attraction between the electric charges of the ions and the electric dipole charges of the water molecules, as illustrated schematically in Fig. 7-6. The same electrostatic attraction exists between the water molecules and the negative charges of the layers. These electrostatic forces cause water uptake between the layers and thereby a reduction of the cohesive forces between them, so that even more water can enter more easily, forcing the layers farther apart. Clays with water can exert quite a considerable swelling pressure. If only monovalent ions are present between the layers, the cohesion may be lost altogether (see Fig. 7-8, 7.2.2.3.1).

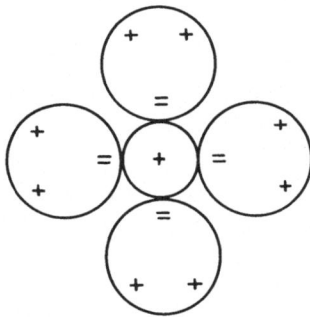

Figure 7-6 Hydration of ions. The distribution of electric charges in water molecules is irregular; each molecule is an electrical quadrupole (indicated by the asymmetrical arrangement of + and −). Therefore that part of the water molecules carrying the negative charge is attracted by the positive charge of a cation, forming an electrostatically bonded water sheath around the ion. The diagram shows an octahedral arrangement (6 water molecules are symmetrically around the ion, the 2 molecules positioned in front and behind the plane of the drawing are not shown). These innermost water molecules around the ion attract additional, less firmly bonded molecules.

In this way the original consolidation process of clay sediments by pressure and increased temperatures may be reversed, resulting in the same water-rich, almost liquid sludge that originally sedimented from the suspension. This process of sedimentation, consolidation (expelling of water), and reversion to the initial state of the sediment is illustrated in Fig. 7-7.

7.2.2.3 Influence of Cations on Water Absorption, Swelling, and Decrease of Shear Strength—Ion Exchange

As already shown above, bivalent and trivalent cations—unlike monovalent cations—effect a stronger cohesion between silicate layers (see Fig. 7-5). Actually the only difference between margarite (brittle mica), which hardly swells at all, and illite, which shows considerable swelling, is that in illite the monovalent potassium ions K^+ effect only compensation of charges between the silicate layers, while in margarite the interposed bivalent Ca^{2+} generates a considerable additional cohesive force. Cations held between the layers by only electrostatic attraction (see Figs. 7-5 and 7-7) are in principle exchangeable against other ions. Such ion exchanges take place in nature and strongly influence strength and cohesion of clay minerals. This is easily demonstrated in the dialyzer, as shown in Fig. 7-3. A thin layer of water-saturated montmorillonite is put on the membrane and the outer space (5) is flushed with a concentrated calcium chloride solution; after some time the montmorillonite will consolidate and decrease in volume, because, following the equation:

$$(Na^+)_2 - \text{montmorillonite} + Ca^{2+} \rightarrow Ca^{2+} - \text{montmorillonite} + 2\,Na^+ \quad (7\text{-}1)$$

sodium ions of the montmorillonite are exchanged against calcium ions from the liquid outside. The driving force of the ion exchange is the tendency of

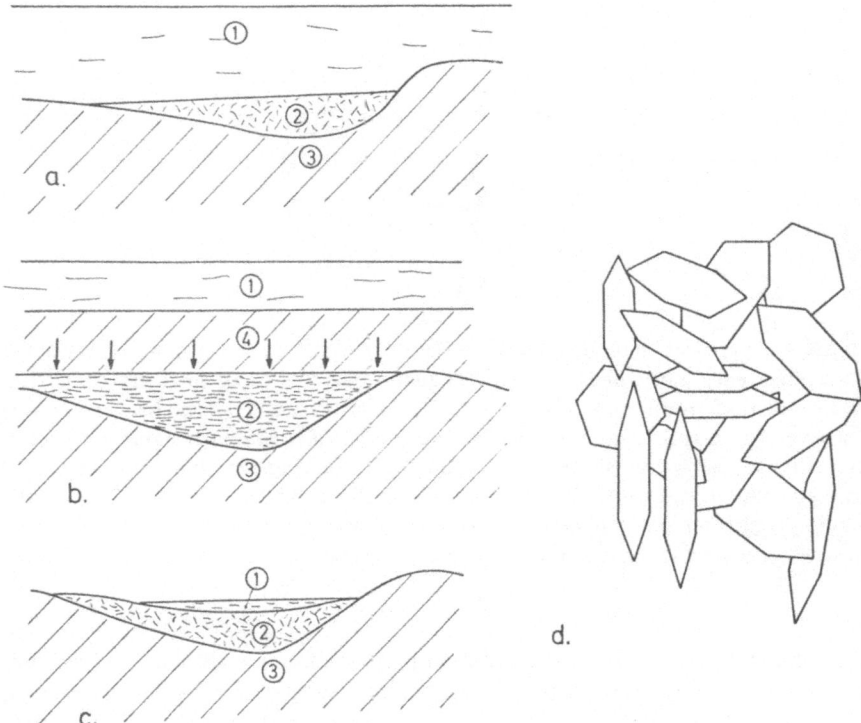

Figure 7-7 Sedimentation, consolidation and reliquefaction of clay soils. a—The watery suspension (1) turns into almost liquid clay sludge (2) which settles into a depression in the solid subsoil. The clay particles are wholly unaligned and surrounded by individual water sheaths; they only touch edge-to-face, water content is high, cohesion low. The structure corresponds to Fig. 7-2 and 7-7 d.

b—During the consolidation process water is pressed out of the clay sludge (2) by the load pressure of superimposed sediments (4); the platelet crystals align themselves face-to-face, thus assuming the least volume. Arrows indicate the direction of pressure from above. Volume and water content decrease sharply, cohesion between particles increases because they are no longer separated by an interposed water sheath and now touch with their faces. The structure corresponds to Fig. 7-1.

c—The superimposed layers are removed (e.g., by erosion), water (1) collects on top of the consolidated clay (2); as it is no longer under pressure, the consolidation process is reversed. Under the influence of hydration energy (see Fig. 7-6), water is again accumulated by the cations between the layers, especially if these are monovalent (Fig. 7-5); distance between particles increases, cohesion decreases, the clay swells and loses its strength. Under unfavorable conditions (e.g., if the penetrating water is rich in sodium and ammonium ions) the initial state of clay sludge can be achieved again, and the structure again corresponds to Fig. 7-2 and 7-7 d.

d—Diagram of structure of thixotropie (highly sensitive) clay. Particles only touch in points and in a moist state the voids are filled with water; resulting low cohesion is destroyed by the least application of force from outside; particles part at the points of contact and the clay turns "quasiliquid." In a dry state this clay is turned into dust by the least mechanical disturbance.

nature to balance an existing difference in concentration: the high concentration of calcium ions in the solution outside causes a diffusion of Ca^{2+} through the membrane into the clay; the high Na^+-concentration in the clay causes a diffusion of Na^+ into the solution outside, where the Na^+-concentration is virtually nil. When the surrounding solution is constantly renewed, so that the Na^+-concentration remains at or near zero and the Ca^{2+}-concentration remains high, then in the end all exchangeable Na^+ ions will have diffused out of the montmorillonite and been replaced by Ca^{2+} ions. The whole process is nothing but a reversion of the plastification of clay and kaoline masses, as practiced in the ceramics industry: to plastify clay or kaoline, it is kneaded with a concentrated soda solution (sodium carbonate solution containing sodium ions Na^+ and carbonate ions CO_3^{2-}), thereby effecting an exchange of Ca^{2+} ions in the material against monovalent sodium ions. In the process the cohesive forces between the silicate layers decrease and the water uptake increases.

7.2.2.3.1   Influence of Road Salting and Sewage Waters on Slides. The influence of the type of cations described above and the possibility of an ion exchange are the reason why contact of clay with water containing higher concentrations of monovalent ions must have an unfavorable effect on the mechanical strength of clay. The process whereby sodium ions are exchanged against the initially present calcium ions is shown in Fig. 7-8.

Salting of roads sometimes produces very concentrated sodium chloride solutions. These solutions may enter the soil on the roadway or at its sides and

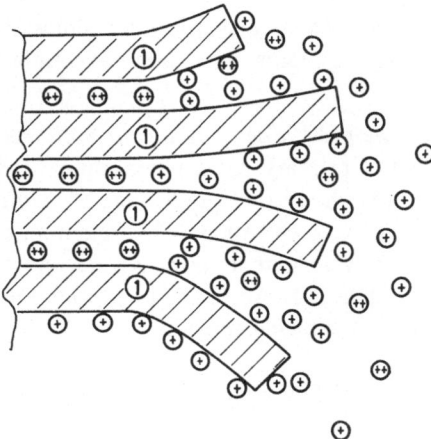

Figure 7-8 Ion exchange of bivalent $(++)$ ions (e.g., calcium ions) against monovalent $(+)$ ions (e.g., sodium or ammonium ions) in a pack of clay mineral layers. Through the high concentration of monovalent ions in the water surrounding the pack, these gradually penetrate between the layers (1) and there displace the initially present, strongly bonding bivalent ions. Owing to loss of cohesion and the hydrate water carried along, the layers are pressed apart and fold out, thus aiding the continuation of the process.

cause an exchange of higher-valency ions against monovalent sodium ions. Under unfavorable conditions this may cause slides due to severe swellings and the resulting decrease of shear strength. Generally, the effects of a saline solution entering in this way will only set in after some time, often after years, because in cohesive soils the water, and with it the ions, will travel very slowly. However, the soil is more permeable for saline solutions than for clear water, as proved experimentally by Casagrande. The reason for this is that an electric potential gradient builds up when water flows through clay soils, which acts in the opposite direction to the flow, and this gradient decreases with a higher conductivity (reversion of electroosmosis, see 7.2.2.4.2). In spite of this, even the flow of saline solutions remains rather slow; only where paths of entry (cracks, crevices, fissures, permeable layers) are already available, can the softening and swelling of the bordering soil progress quickly. In some cases a fast penetration also takes place at the zone of contact between layers of different soils, even if both are cohesive (see 3.3.3). Where such paths of entry are not available, the maximum velocity of penetration in cohesive soils is several millimeters per day or less. Even when no slides occur, roads suffer great damage from the concentrated saline solutions produced by winter salting. The penetrating saline solutions cause increased swelling and softening and this destroys the road paving in places, which in turn aids further penetration of the saline solution.

The same process may be triggered by sewage and industrial waste waters. Sewage from households and stables contains the monovalent ammonium ion NH_4^+ in large quantities and also large quantities of common salt. Discharges from dung heaps and pools of liquid manure as well as seeping and trickling household sewage may be factors responsible for slides. The same applies to industrial waste water, if it contains monovalent ions in larger quantities.

7.2.2.3.2 Influence of the pH-Value of Water in Soil.

As already stated, the chemical decomposition of primary rocks is mainly caused by acids dissolved in water. This fact already shows the significance of acidity or basicity of the pore water in the soil. The concentration of hydrogen ions c_{H^+} is the deciding factor for whether and to what extent water is acid; these ions are the active agent of all acids. Hydrogen ions are formed from all acid molecules by way of dissociation in solution. Following the equation

$$CH_3COOH \xrightarrow{H_2O} H^+ + CH_3COO^-$$ (7-2)

hydrogen ions H^+ and acetate ions CH_3COO^- are formed from acetic acid CH_3COOH. Hydrogen ions cause the sour taste and all other acid properties of vinegar. For example, the dissolution of lime in acetic acid is due to the hydrogen ion present. However, for the processes in nature, carbonic acid is of much greater importance. Carbon dioxide, CO_2, which forms carbonic acid when dissolved in water, is present everywhere in the air and in many parts produced by volcanic activities. By itself it is not acid, as it does not contain hydrogen which could be set free as hydrogen ions, H^+. Carbon dioxide will, however, dissolve in water and form carbonic acid following the equation

$$CO_2 + H_2O \rightarrow H_2CO_3, \tag{7-3}$$

from which two H^+ can be set free:

$$H_2CO_3 \overset{H_2O}{\rightarrow} H^+ + HCO_3^- \tag{7-4}$$

$$HCO_3^- \overset{H_2O}{\rightarrow} H^+ + CO_3^{2-} \tag{7-5}$$

whereby hydrocarbonate ions HCO_3^- and carbonate ions CO_3^{2-} are formed.

Carbonic acid is a weak acid, meaning that only a few acid molecules will dissociate to produce hydrogen ions following the above equation; most of the acid molecules will retain their hydrogen in the solution. Furthermore, water will dissolve only relatively little carbonic acid, but still enough so that any water that has been in contact with air over a certain period will show easily discernible acid properties and will be capable of dissolving limestone (the formation of soluble complexes between calcium ions and hydrocarbonate ions contributes to this process). The chemical decomposition of rock during weathering is caused mainly by the dissolving action of water containing carbonic acid.

Just as the presence of hydrogen ions H^+ makes water acid, the presence of hydroxyl ions OH^- makes it basic: the active agent of all bases is the hydroxyl ion. When sodium lye NaOH dissolves in water, sodium and hydroxyl ions are formed:

$$NaOH \overset{H_2O}{\rightarrow} Na^+ + OH^- \tag{7-6}$$

Since sodium lye is a strong base, virtually all particles dissociate following the above equation; the concentration of hydroxyl ions equals that of dissolved sodium lye. Just as there are weak acids, there are also weak bases from which only part of the molecules will dissociate.

The pH-value is the standard measure for the concentrations of hydrogen ions as well as hydroxyl ions and thereby for acidity or basicity, even though it is only a logarithmic function of the hydrogen ion concentration:

$$pH = -\log c_{H^+} \tag{7-7}$$

or

$$c_{H^+} + 10^{-pH} \tag{7-8}$$

The pH-value may be used to characterize the degree of basicity and of acidity, because the product of hydrogen ion concentration and hydroxyl ion concentration in aqueous solutions is approximately a constant which depends only upon temperature:

$$c_{H^+} \cdot c_{OH^-} = 10^{-14} \quad \text{(at } 25°C) \tag{7-9}$$

from which results

$$pH - \log c_{OH^-} = 14 \qquad (7\text{-}10)$$

and

$$c_{OH^-} = 10^{pH-14}. \qquad (7\text{-}11)$$

In these equations the concentrations are in moles and are not as low as could be concluded from the figures. One mole, after all, consists of $6{,}023 \cdot 10^{23}$ particles of the same kind, meaning that at a hydrogen ion concentration of only 10^{-5} moles/liter, there are still $6{,}023 \cdot 10^{23} \cdot 10^{-5} = 6{,}023 \cdot 10^{18}$ hydrogen ions in one liter.

The *neutrality point* divides the range of acidity from that of basicity. At this point the concentrations of hydrogen ions and hydroxyl ions are equal by definition. Accordingly, the relation $c_{H^+} \cdot c_{OH^-} = 10^{-14}$ in a neutral aqueous solution gives

$$c_H + = c_{OH^-} = 10^{-7} \text{ mol/liter} \qquad (7\text{-}12)$$

and the pH-value at the point of neutrality equals 7.

All solutions with a pH-value lower than 7 (hydrogen ion concentrations higher than 10^{-7} moles/liter) are considered acid; all solutions with a pH-value higher than 7 (hydroxyl ion concentration higher than 10^{-7} moles/liter) are called basic (for determination of pH-values see 5.1.8.2). The pH-value, i.e., the hydrogen ion concentration, of water in contact with clay minerals plays a significant part in the properties of clayey soils.

As previously stated, acid waters, and here especially water containing carbonic acid, cause the decomposition of minerals by weathering. But since these weathering processes extend over geological periods, they are of no immediate interest to the geotechnical engineer. Much faster, however, is the process of simple ion exchange, whereby hydrogen ions are exchanged between soil and pore water. Besides that, the pH-value is also a deciding factor for the redox-potential of the soil and thereby for the electroosmotic transport of water that may generate slip surfaces (7.2.2.4 and 7.2.2.5).

If monovalent alkali cations (K^+, Na^+) are replaced by hydrogen ions, water uptake and swelling are reduced because hydrogen ions are bonded to the silicate layers not only by electrostatic forces but also chemically. Therefore, silicate layers, or layered packs, which otherwise carry a strong negative charge, become less charged or carry almost no charge, and this reduces their tendency to take up water. On the other hand, the cohesive forces between silicate layers are reduced by hydrogen-ion uptake and the physical properties come close to those of nonswelling layered silicates, such as talcum or pyrophyllite, which have a rather low shear strength without taking up water and swelling. Especially in comparison to clay minerals containing bivalent and trivalent cations, shear strength is reduced at equal water content. However, since natural waters usually have a low acidity (this applies also to most of the acid waste waters), ion exchange against hydrogen ions is of little practical significance. The important exception to this rule takes place in electroosmotic drainage, where such an ion exchange at the anode may be of significance (see 7.2.2.4.2 and 6.1.5.2).

Much more important is the influence of high pH-values. Many waste waters, in particular sewage containing feces, are strongly alkaline. The hydroxyl ions that they contain bond chemically to clay particles and increase their negative charge. This higher negative charge increases the quantity of cations required to compensate it, and thereby the quantity of hydrated water increases too, especially if the water contains high concentrations of strongly hydrated monovalent ions (Na^+, NH_4^+, etc.), as is mostly the case in such alkaline waste waters. In addition, the high pH-value causes a deactivation of many high-valency cations like Mg^{++}, Fe^{++}, Fe^{+++} and partially also Ca^{++}, by precipitating them as hydroxides and carbonates. This contributes somewhat to the consolidation of the soil, as they cause a mechanical cementing; on the other hand, they are no longer available to compensate electric charges. Therefore, monovalent cations have to take their place, which attract more water and bond the layers less firmly. In summary, alkaline solutions effect stronger swelling and lower shear strength.

Another unfavorable effect of high pH-value is a shift of the electric soil potential towards negative values (see 7.2.2.5), which may generate an additional *water inflow by electroosmosis* (see 7.2.2.4), causing a further decrease of shear strength.

Quite generally, any significant deviation from the normal soil pH (about 6.0 to 8.5) has an unfavorable effect upon shear strength and water uptake to clay minerals, with upward deviations being much more harmful.

7.2.2.3.3 Thixotropy—Quick Clay The most dangerous landslides in clay soils are caused by thixotropic behavior. In a system of clay minerals and water there is a labile state of strength which may be disturbed by minor mechanical influences that result in a reduction of shear strength nearly to zero. Sométimes a whole soil formation will virtually liquefy and flow over long distances, even if the slope of the terrain is small (Bjerrum, 1954).

This phenomenon may be caused by a purely mechanical increase of porewater pressure in water-saturated loose soils (sand); or in dry loess soils by excess air pressure in the pores, which is sometimes produced by earthquakes.

Very dangerous liquefaction slides also occur in thixotropic clay soils (so-called *quick clay*), which are for the most part recent marine sediments with a rather high salt content. In such soils the liquefaction may be triggered by very minor mechanical effects (construction work, heavy transports, heavy buildings), and after such a slide has started it will propagate itself over long distances. The transgression from a (relatively) solid to a quasiliquid state requires only very little energy. In rheology this phenomenon has long been known and is called *thixotropy*.

In quick clay, thixotropic liquefaction is easily understood from the morphological structure of clay minerals. The primary sedimentation of clay minerals from suspension at first produces a quasiliquid sludge of very low shear strength and high water content. In this sludge the platelet particles of clay minerals are completely unaligned and held together only by very low bonding forces, because each clay particle has its own water sheath (Figs. 7-7a and d).

As already described the consolidation of this quasiliquid sludge may take place under the pressure of superimposed soil layers; high pressure over a long period of time transforms it to consolidated clay with particles aligned face-to-face, as shown in Fig. 7-7b.

Without consolidation pressure, the sludge consolidates of its own, simply by the clay particles' partially uniting their individual water sheaths. This produces cohesion by way of the positive cations contained in the water. There is no parallel alignment of the particles because there is no pressure to enforce it; water content and water-uptake capacity are reduced, but not nearly as much as by consolidation under pressure. Cohesion is low because the particles are bonded only where they touch by points and edges, not with the much larger area of their faces. If the positive cations are mainly monovalent Na^+ ions, as in marine quick clay, cohesive forces are especially low.

From this it is obvious that the "house-of-cards" structure of such quick clay, as shown in Fig. 7-7d, will collapse under minor mechanical stress and that a slide, once it has started in one place, may liquefy bordering areas. Besides that, the water uptake in such clays is fast on account of its structure, and heavy rainfalls may cause liquefaction even without the influence of mechanical stress.

The shear behavior of such clays is shown in Fig. 7-9, corresponding to a combination of Bingham-type behavior with low initial tension and purely thixotropic behavior. The curve is easily explained by the structure of quick clay.

7.2.2.4 Electric Charge of Clay Particles and Related Phenomena
The crystal layers of clay minerals carry negative electric charges which are compensated by positive cations (Figs. 7-4 and 7-5). Therefore, even larger

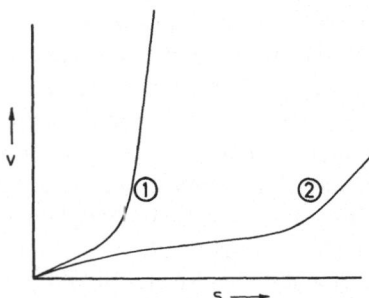

Figure 7-9 Shear behavior of thixotropic clay (1) and common soils (2); $v =$ shear velocity; $s =$ shear stress. Thixotropic material loses its strength completely and turns quasiliquid by application of a low stress. The behavior of common soils is an initial consolidation and, with much higher stress, plastic behavior with considerably higher strain resistance.

clay particles consisting of many such layers carry a negative charge, for which surrounding positive cations have to compensate to maintain electric neutrality.

These adsorbed cations are themselves surrounded by a water sheath (they are hydrated; Fig. 7-6), and therefore the electrostatic attraction is not strong enough to bond the cations to the surface as a rigid layer. In some cases there is even not enough surface area for all the ions needed to compensate the electric charge. As a consequence, a statistical distribution of positive ions collects near the particle which produces a space charge in this region; outside, the water is electrically neutral and not influenced by the charged particle ("free water"), as shown in Fig. 7-10. All the water beyond a distance of more than several 0.01μm is to be considered free water; this means that the following conclusions are valid not only for water around suspended or colloidally dissolved clay particles, but also for water contained in microscopic and submicroscopic soil pores.

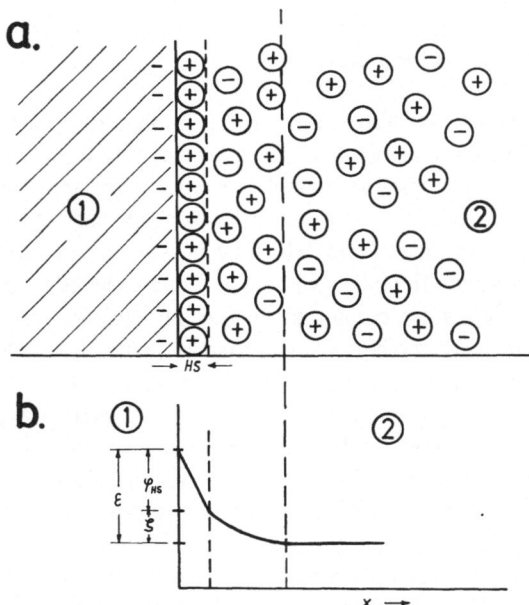

Figure 7-10   Electrochemical double layer. a—The layer of hydrated positive ions (the circles symbolize positive or negative ions, including their water sheaths) is firmly bonded to clay particle (1) with negative charge. In this so-called Helmholtz layer (*HS*) the major part of the potential difference φ_{HS} between clay particle and solution becomes almost linear. In the adjacent zone the concentration of positive ions decreases gradually (following an exponential function) to the average value in the solution (2). The potential drop in this "diffuse layer" is the electrokinetic or ζ-potential. The layer still carries a positive charge but is not rigidly bonded to the surface. b—Electric potential profile in relation to the distance from the surface χ; ϵ is the total potential difference between solid and solution.

7.2.2.4.1　Electrochemical Double-Layer. Corresponding to Fig. 7-10, the layer of hydrated ions accumulated in the vicinity of a charged surface may be considered formally as consisting of two separate parts: an inner, rigidly bonded layer (reaching in Fig. 7-10 to the dashed line), called the Helmholtz layer, and an outer layer in which an excess concentration of positive ions gradually decreases to the average concentration in the water phase. This creates a distribution of the electric potential as shown in Fig. 7-10b. The thickness of the whole layer depends upon the type of cations. Cations with small diameters have a higher electric field, more strongly attract the electric dipoles of water molecules, and are therefore more strongly hydrated; they have a thicker water sheath. Consequently, *the electrochemical double layer*, in which the electric potential decreases from the negative value at the surface of the particles to the more positive value in the solution, will be thicker with smaller ions than with larger ones. The thickness of the layer increases in the sequence $Cs^+ <$ $Rb^+ < NH_4^+ < K^+ < Na^+ < Li^+$. The absolute thickness of the layers is in the range of approximately 0.005 μm to 0.3 μm, the highest value being determined for bentonite with sodium ions. Divalent cations (Ca^{2+}, Mg^{2+}, Ba^{2+}, Sr^{2+}) generally form significantly thinner layers on account of the strong bonding to the surface of clay minerals.

7.2.2.4.2　Electroosmosis. One of the most significant results of the double-layer theory is that the ions of the outer layer are only loosely bonded and able to move rather freely, while the ions of the inner layer are more or less firmly bonded to the solid surface. This outer layer of water carries an electric charge and is able to move in an electric field. The positive ions of this layer move in an electric field towards the negative pole and carry the adhered hydrate water along. If the clay particles cannot move—as in solid soil—only the water in this layer, together with the cations, will migrate, causing an electrolytic transport of water to the negative pole. This flow of water, in our case towards the negative pole, is called *electroosmosis* (if the solid carries a positive charge, water transport to the positive pole would result). If the colloidal solids with negative charge are not fixed but mobile (colloidal solutions or suspensions), they will migrate in an electric field towards the positive electrode, while the positive cations migrate to the negative electrode as before. In this way, ion-deficient clay particles are collected, and at the same time dehydrated at the positive electrode. The phenomenon is used for electrophoresis purification of kaoline and clay.

Transport of water through fine capillaries is obviously slowed down by opposing forces: On the one side by the viscosity of water, because the layer immediately at the surface of the capillary wall remains in place and therefore an inner friction develops within the liquid; on the other side by electrostatic attraction between the ions and the solid surface carrying an opposite charge, which tends to hold the ions in place. Owing to the shielding effect of the rigid layer (Fig. 7-10), however, this attraction force is low and may be neglected in the calculations.

When the flow of liquid from the positive to the negative pole in a capillary

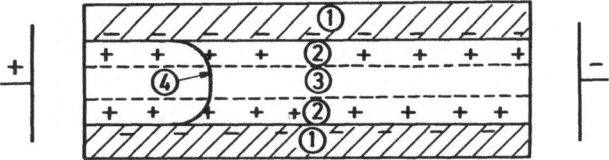

Figure 7-11 Capillary with electroosmotic flow; (1) wall of capillary (negatively charged); (2) mobile liquid layer carrying excess positive charge (diffuse layer, see Fig. 7-10); (3) uncharged capillary water being moved along only by force of friction; (4) profile of flow velocity.

tube is stationary (Fig. 7-11), then the driving electrostatic force is just compensated by the effect of the internal friction of the liquid, that is,

$$E \cdot \rho = \eta \frac{d^2 v}{dx^2}. \tag{7-13}$$

(E = electric field strength; ρ = charge density; η = viscosity of water; v = stationary velocity of liquid at a distance x from the surface.) By converting the charge density ρ with the aid of Poisson's equation to the electric potential ψ, and considering that ψ as well as v depend only upon the vertical distance from the surface of the capillary wall, one arrives at

$$\frac{d^2 v}{dx^2} = \frac{\epsilon \epsilon_0 E}{4 \pi \eta} \cdot \frac{d^2 \psi}{dx^2} \tag{7-14}$$

for a liquid with dielectric constant ϵ (ϵ_0 = vacuum dielectric constant.)

Integrating this equation from $x = 0$ to $x = \infty$, and with the boundary conditions that at $x = \infty$, $\psi = 0$, $d\psi/dx = 0$ and $dv/dx = 0$, one arrives at the Helmholtz-Smochulowski equation for stationary flow velocity in electroosmosis

$$v = \frac{\epsilon \epsilon_0 E \zeta}{4 \pi \eta}, \tag{7-15}$$

ζ (the electrokinetic potential) being the electric potential ψ at the plane between the rigid, immovable layer and the mobile liquid (this plane of division is marked in Fig. 7-10 by a dashed line). v is the stationary flow velocity outside the electrochemical double-layer, where ψ already equals 0. Consequently, this equation is only valid for capillaries with radii significantly larger than the thickness of the electrochemical double layer; for very fine capillaries there are deviations (Kortüm, 1966).

Absolute values of the electrokinetic potential are quite small; depending upon type and concentration of the positive counter-ions, they are between 0.01 and 0.1 V.

Experimentally, v is calculated from the volume of liquid transported per second (electroosmotic penetrability D)

$$D = v \pi r^2 = \frac{\epsilon \epsilon_0 \zeta r^2}{4 \eta} E \ [\text{m}^3/\text{sec}] \tag{7-16}$$

for a single capillary with a radius r. For a porous solid body one gets

$$D = v \sum_1^i \pi r_i^2 = \frac{\epsilon \epsilon_0 \zeta \sum_1^i r_i^2}{4\eta} E \ [m^3/sec] \qquad (7\text{-}17)$$

with all constants and measured values in the units meters, seconds, and volts.

The equation indicates that the volume of water transported is proportional to electric field strength $E \ [V \cdot m^{-1}]$ and to the square of the pore radii, but is independent of the length of the capillaries.

Simultaneously with the transport of water, electric current flows through the capillary or the capillary system, owing to the migrating positive ions. According to Ohm's law

$$I = \frac{U}{R} = \frac{E \cdot l}{R} = \kappa \pi r^2 E, \qquad (7\text{-}18)$$

where κ is the specific conductivity of the liquid. By combination with the preceding equation one gets

$$D = \frac{\epsilon \epsilon_0 \zeta I}{4\pi \eta \kappa}, \qquad (7\text{-}19)$$

which means that D can be calculated from conductivity κ of the pore liquid—which is easily determined by experiment—and the measured d.c. current I.

Also of importance for the subject is the *electroosmotic pressure* developing between the two ends of a capillary system when an electric tension is applied. If a d.c. current is applied to the electrodes in an apparatus as shown in Fig. 7-12, water flows to the negative electrode; here the water table rises while it sinks on the side of the positive electrode. However, after a while the water tables will become stationary because a return flow through the capillaries sets in under the influence of the pressure difference Δp (given by $\Delta h \cdot \rho \cdot g$, ρ = density $[kg \cdot m^{-3}]$ of the liquid; g = gravity constant, i.e., for water $\Delta h \cong$

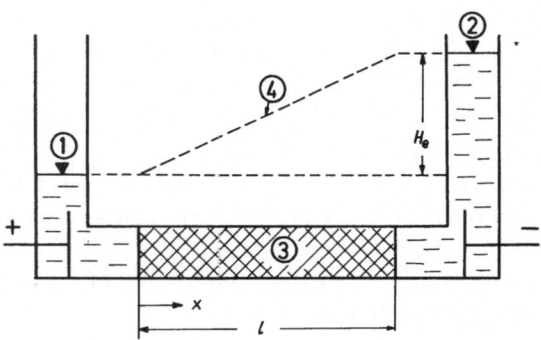

Figure 7-12 Apparatus for measuring electroosmotic height H_e; (1) water table, maintained constant at anode; (2) water table at cathode, after reaching maximum height H_e; (3) soil sample; (4) profile of pore pressure versus distance χ, from 0 to 1.

$\Delta p/g$). This return flow follows Poiseuille's law for laminary flow

$$D'v' = \frac{\Delta p \pi r^4}{8\eta l} \text{ [m}^3/\text{sec]} \tag{7-20}$$

(l = length of capillary). The stationary state and thereby the highest possible difference in pressure (or maximum electroosmotic height) is reached when electroosmotic flow and return flow velocity are equal ($v = v'$) and

$$\Delta p = \frac{2\epsilon\epsilon_0 \zeta l}{\pi r^2} E = \frac{2\epsilon\epsilon_0 \zeta l}{\pi^2 r^4 \kappa} I. \tag{7-21}$$

This means that electroosmotic pressure and electroosmotic height should be proportional to the applied tension U, but independent of the length of the capillaries (since $E = U/l$, the length of the capillaries is cut out). In the experiment shown in Fig. 7-12, the length of the porous solid has no influence upon the electroosmotic height H_e.

Electroosmosis is of practical interest for the drainage of cohesive soils, as it reduces the susceptibility to slides (Casagrande method; see 6.1.5.2). Furthermore, it is possible to put the water contained in clay soils to use by "electric pumping", meaning that capillary water may be driven into wells by application of electric energy between a negative electrode mounted in the well and several positive electrodes around the well. This water is not obtainable by other methods since it is held back by capillary forces. On the other hand, water may accumulate in certain soil layers by natural electroosmosis, reducing shear strength and generating slip surfaces causing landslides.

It is common knowledge that ground water rises up in soil pores by capillary action. This rise against the force of gravity is the consequence of the low surface tension between water and soil and follows the equation

$$h = \frac{2\sigma}{g \cdot \rho \cdot r} \tag{7-22}$$

(h = height of capillary rise; σ = surface tension of water; g = gravity constant; ρ = density of water; r = radius of capillary). In soils with a very tight capillary system the capillary rise may be quite considerable. For a capillary radius of 1 μm, $\sigma = 73 \cdot 10^{-3}$ N/m and $\rho = 10^3$ kg/m^3; the resulting rise is

$$h = \frac{2 \cdot 73 \cdot 10^{-3}}{9.81 \cdot 10^3 \cdot 10^{-6}} = 14.9 \text{ m} \tag{7-23}$$

Natural electroosmosis due to electric potential differences in the soil may also cause the ground water to rise. However, the maximal electroosmotic height H_e of such a natural electroosmosis is far less than the possible capillary rise in the soil. Electric tensions caused by the different chemical compositions of the various soil layers are of the magnitude of 0.03 to 0.1 V, in exceptional cases 0.2 or 0.3 V. Even for a capillary radius of 0.1 μm and a tension of 0.3 V the calculated electroosmotic height is

$$p = \frac{V\rho g}{\pi r^2} = hg\rho = \frac{8\zeta\epsilon\epsilon_0 U}{r^2} \tag{7-24}$$

$$h = \frac{8 \zeta \epsilon \epsilon_0 U}{g \rho r^2} = \frac{8 \cdot 0.1 \cdot 80 \cdot 8.85 \cdot 10^{-12} \cdot 0.3}{9.81 \cdot 10^3 \cdot (10^{-7})^2} = 1.73 \text{ m} \qquad (7\text{-}25)$$

Of course, the above equation predicts greater electroosmotic heights with a further decrease of pore radius, but in very small capillaries the electroosmotic height becomes virtually independent of the pore radius (Schmid, 1951).

Therefore, the significance of natural electroosmosis for landslides is not the transportation of ground water to great heights. The important effect is an accumulation of water in narrowly defined areas, especially *at the boundary between reducing soils (with a negative electric potential) and oxidizing soils (with a positive electric potential). Contrary to capillary rise, in electroosmosis the pore liquid is under marked capillary excess pressures. Therefore, water can even be absorbed against the pressure of superimposed soil and cause swelling and reduction of shear strength; electroosmosis drives water from oxidizing soil into the boundary layer to reducing soil and generates a slip surface there* (Fig. 7-13).

Electroosmotic drainage produces quite remarkable rise heights or underpressures because the applied d.c. voltages are quite high; up to and over 100 V (6.1.5.2). For practical applications, the electroosmotic water transport is derived from two empirical soil coefficients. As previously outlined, electroosmotic water transport or electroosmotic height results from the interaction of two opposing effects: The backflow, which is generated by the pressure of electroosmotic rise height, counteracts the flow caused by the electric field. The backflow is empirically defined by Darcy's law, which is valid for hydraulic ground-water flows,

$$v_h = - k_h \cdot \text{grad } H \qquad (7\text{-}26)$$

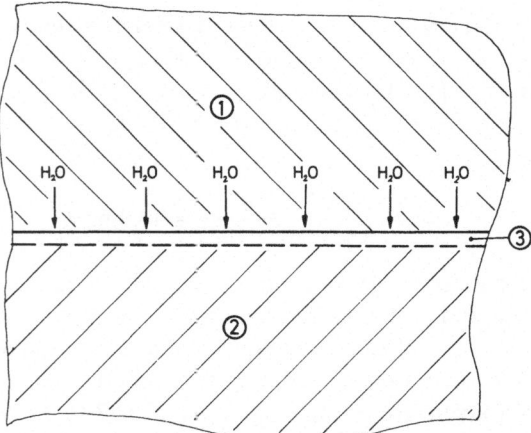

Figure 7-13 Natural electroosmotic flow of water from an oxidizing soil layer (1) to the zone of division (3), marked by a dashed line, and to a reducing soil layer. In (3) the electroosmotic pore-pressure may cause swelling and loss of strength, even against an existing consolidation pressure, and generate a slip surface.

In this equation v_h is hydraulic flow rate caused by the existing gradient of hydraulic height H; k_h = hydraulic soil coefficient = velocity of flow at a gradient of 1, i.e., under pressure of a 1 m high water column in a soil layer of 1 m thickness, in the SI measuring system; H is to be expressed in meters and v_h and k_h in m/sec. This relation is the three-dimensional equivalent to the one-dimensional Poiseuille equation (7-20) already presented.

Just as the hydraulic properties may be characterized by the empirical quantity k_h, the electroosmotic properties may by characterized by the electroosmotic coefficient k_e. Using k_e, the electroosmotic flow rate v_e will correspond to Eq. (7-15) and becomes analogous to Darcy's relation (7-26)

$$v_e = -k_e \cdot \text{grad } U \tag{7-27}$$

or

$$v_e = k_e \cdot E \tag{7-28}$$

where k_e is the electroosmotic flow rate at a field intensity $E = 1$ V/m and has the dimension $[\text{m}^2/\text{V} \cdot \text{sec}]$. If both types of flow take place simultaneously, as is usual in practical cases, then

$$v = v_h + v_e \tag{7-29}$$

and therefore

$$v = -k_h \text{ grad } H - k_e \text{ grad } U \tag{7-30}$$

$$v = -k_h \text{ grad } \left(H + \frac{k_e}{k_h} U \right) \tag{7-31}$$

The maximum electroosmotic rise height may be calculated just as Δp was in Eq. (7-21), taking into account that at maximum rise height H_e, the upward flow rate v_e must be equal to the hydraulic downward flow rate v_h caused by height H_e, which means that equilibrium between electrical and gravitational forces is reached. For the one-dimensional case,

$$v_e = k_e \cdot \frac{U}{l}$$

and therefore

$$v_h = k_h \cdot \frac{H_e}{l} \tag{7-32}$$

(l = length of soil sample, e.g., in an apparatus as shown in Fig. 7-12). From this follows

$$H_e = \frac{k_e}{k_h} \cdot U \tag{7-33}$$

For $U = 1$ V another empirical coefficient is defined—the specific electroosmotic rise height

$$h_e = \frac{k_e}{k_h} \tag{7-34}$$

h_e has the dimension [m/V] and indicates if substantial electroosmotic water transport is possible in the investigated soil. A condition for significant electroosmotic water transport is the possibility of a significant electroosmotic rise height, which according to Eq. (7-33) evidently only exists if k_e/k_h is not too small. This in turn means that the radii of the pores must be sufficiently small, because only in small pores is electroosmotic flow at an advantage versus hydraulic flow. Usually this criterion is stated as

$$h_e = \frac{k_e}{k_h} \geqslant 0.1 \; [\text{m/V}]. \tag{7-35}$$

If (7-35) applies to a soil, electroosmotic drainage can be applied successfully. According to available investigations, k_e varies in cohesive soils only slightly, the average value being $k_e = 5 \cdot 10^{-9}$ m^2/V $\cdot$ sec (the highest variations are registered in bentonite, for which—depending upon water content—the relation between highest and lowest value is 15:1). Therefore, primarily the k_h-value determines the possibility of successful application of electroosmotic drainage or the possibility of natural electroosmosis, since k_h-values of different soils vary by a factor of 10^5. Moreover, k_h may be reduced in a soil to one-tenth by mechanical consolidation.

Electroosmotic pore-water pressure only exists if $H_e > 0$, i.e., if the water table or the zero-pressure plane at the cathode rises above the initial value. If at the cathode the water table is kept constant (for example by an overflow or by pumping during electroosmotic drainage), and if the anode water table remains constant too, the friction force of the moving water is at equilibrium with the driving force of the electric field and no additional pore-water pressure is generated. A situation where $H_e = 0$ hardly ever occurs in practice, neither in natural electroosmosis nor when applying electroosmotic drainage. In the apparatus shown in Fig. 7-12, the state $H_e = 0$ can be easily achieved by keeping the water tables on the right and on the left at the same height. But even by changing the arrangement of the electrodes, as in Fig. 7-14, one gets a pore-

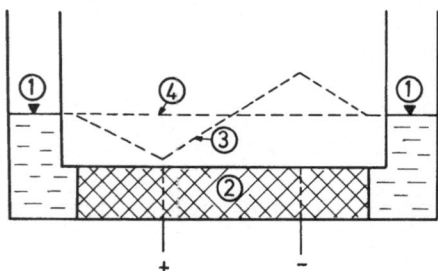

Figure 7-14 Change of pore-water pressure by electroosmosis, if water flow to the anode ($+$) and from the cathode ($-$) is constricted; (1) water table kept constant at anode and cathode; (2) soil sample, ($+$) and ($-$) = penetrable anode or cathode (net); (3) profile of pore pressure with electroosmosis; (4) profile of pore pressure without electroosmosis. In spite of an equal hydraulic pressure on both sides, negative pressure develops at the anode and overpressure at the cathode.

water pressure profile as illustrated, even though the hydraulic pressure is equal on both sides of the soil sample.

Therefore, in real electroosmosis there are normally positive and negative pore-water pressures. In the quasi–one-dimensional case of the apparatus in Fig. 7-12, the calculation of pore-water pressures is simple:

$$\frac{\Delta p}{dx} = \rho_{H_2O} \cdot g \frac{dH_e}{dx}$$

$$p(x) = \int_0^x \rho_{H_2O} \cdot g \frac{dH_e}{dx} \cdot dx = \rho_{H_2O} \cdot g \frac{H_e}{l} \cdot x \qquad (7\text{-}36)$$

$$p(x) = 10^3 \cdot 9.81 \frac{H_e}{l} \cdot x[\text{N/m}^2]. \qquad (7\text{-}37)$$

The pore-water pressure p is not necessarily a positive quantity; under appropriate conditions negative pressures may be produced. For example, in Fig. 7-14, pressure at the anode is considerably decreased by electroosmosis. If the tube on the side of the anode is closed tightly, no water flow is possible. The consequence is a considerable negative pressure in the pores at the anode, which consolidates the soil sample in this area, while an excess pressure is maintained at the cathode which may cause swelling.

In this connection it is of importance that with greatly restricted water flow to the anode the soil material may dry up, which causes an increase of the electric resistance of the soil around the anode. In practical applications of electroosmotic drainage this drying-up is accelerated by the heat generated by the current flow. Consequently, in electroosmotic drainage the electric power applied has to be low enough to allow sufficient replenishment of water at the anode by capillary rise. Empirically, the temperature around the anode must not increase by more than 5°C.

7.2.2.5 Electric Soil Potentials, Reducing and Oxidizing Soils, Correlation between Soil Potentials and Landslides

As outlined in 7.2.2.4, *natural electric potential differences present in the soil* may cause an electroosmotic transport of water, which in turn may cause an increase of water pressure and a decrease of shear strength in certain layers, and consequently produce landslides. Evidently such differences in electric potential can only be caused by differing chemical properties of different parts of the soil, and can only be *maintained by processes which continuously provide the required energy.* If there is an electric potential difference and a finite resistance, a current flow ensues tending to eliminate this difference. The electric resistance of soils is usually high, but not infinite; therefore there is always a current flow between areas with different electric potentials (which may cause a transport of water). *The energy $I \cdot U$ for this current has to be supplied by chemical or mechanical energy.*

The existence of electric potential differences (self potentials—SP) between different types of soil layers has been proved experimentally beyond any doubt (for methods of measurement, see 5.1.8.1). Therefore, the generation and effects of these electric tensions in the soil remain to be investigated.

The basic difference between the types of soil is their content of oxidizing and reducing substances (Veder, 1963). Everybody working with soil is acquainted with strongly reducing soils, such as putrid mud, marshy soils, or blue clay; their reducing properties usually can be quickly recognized by the rotten smell. This smell disappears if the soil is in contact with air for some time because it is oxidized by the oxygen in the air. All reducing soils have a dark to blue color, caused partly by their content of organic substances and partly by sulfides and oxide hydrates of bivalent iron. However, oxidizing soils are yellow to reddish brown, owing to oxide hydrates of trivalent iron.

In air, the reducing substances of the soil are oxidized by oxygen:

$$4 \, (FeO \cdot H_2O) + O_2 \rightarrow 4 \, (FeO \cdot OH) + 2 \, H_2O, \tag{7-38}$$

$$FeS + 6 \, H_2O + 9 \, O_2 \rightarrow 4 \, (FeO \cdot OH) + 4 \, SO_4^{2-} + 8 \, H^+, \tag{7-39}$$

which converts iron(II)oxidehydrate to iron(III)oxidehydrate, and iron(II)sulfide to iron(III)oxidehydrate, sulfite ions, and hydrogen ions. Both reactions are reversed if iron(III)hydrate and sulfate are reduced by organic matter (often with participation of bacteria). The reverse reactions are not exactly identical, as oxygen is not released as a gas but reacts with organic substances to form the final products of carbonic acid and water. In fact, this back-reaction cannot be stated in exact equations; for the subject at hand, (7-38) and (7-39) sufficiently illustrate the principle. From these equations it may be seen that the energy required to maintain electric potential differences is mainly produced by two chemical reactions: On one side, the reaction between oxygen from the air and reducing soil components and on the other side the reaction between reducing substances or anaerobic soil bacteria with iron(III)oxidehydrate and sulfate. Both reactions tend to maintain different concentrations, particularly of iron(II) and iron(III) compounds, in the different soil layers. Assuming an abundance of iron compounds to be in balance with all other substances, the concentration of iron(II) and iron(III) ions has to be considered the determining factor for reducing or oxidizing soil properties.

With known concentrations of iron(III) and iron(II) ions, the local electric potential of the soil can be calculated from Nernst's equation

$$U = 0.771 + \frac{RT}{F} \ln \frac{c_{Fe^{3+}}}{c_{Fe^{2+}}} \tag{7-40}$$

(0.771 = standard potential of the reaction $Fe^{3+} + e^- \rightleftarrows Fe^{2+}$; R = gas constant = 8.31 J/K·mol; T = absolute temperature in K; F = Faraday constant = 96490C; $c_{Fe^{3+}}$ and $c_{Fe^{2+}}$ = concentrations of iron(III) and iron(II) ions in mol/l of the pore water. The total concentrations of iron(III) and iron(II) compounds in the soil cannot be used for these calculations, since only dissolved substances are electrochemically active. Therefore, the concentrations in the pore water must be used, which are more difficult to determine. The electric tension U between points 1 and 2 may be calculated from two different values from Eq. (7-40)

$$U_{1,2} = U_1 - U_2 = \frac{RT}{F}\left(\ln \frac{c_{Fe^{3+},1}}{c_{Fe^{2+},1}} - \ln \frac{c_{Fe^{3+},2}}{c_{Fe^{2+},2}} \right) \qquad (7-41)$$

Calculated values from Eq. 7-41 are always higher than experimental results since, first, Eq. (7-41) is strictly valid only if there is no electric current flow (resistance $= \infty$), and second, since the flowing current creates an additional diffusion potential in the opposite direction.

The origin and magnitude of this electric tension can be easily demonstrated experimentally. If in the arrangement shown in Fig. 7-15 the left compartment is filled with a solution of iron(III) and iron(II) ions in the concentration ratio 1:100 (reducing solution corresponding to Eq. (7-40)), and the right compartment is filled with a solution also containing iron(III) and iron(II) ions, but in a ratio of 100:1 (oxidizing solution corresponding to Eq. (7-40)), the pH-meter

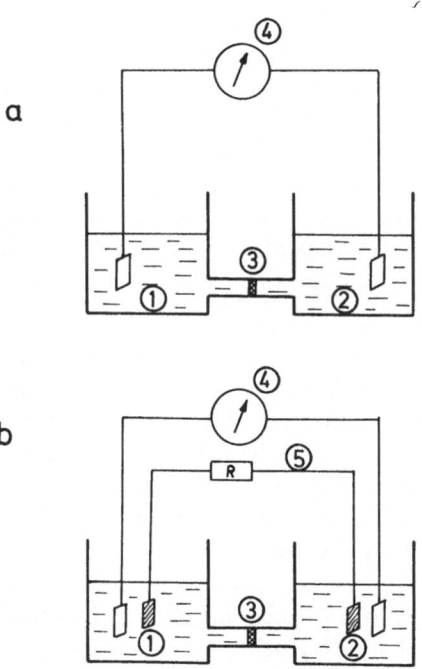

Figure 7-15 Demonstration of origin and magnitude of soil redox potentials. a—Two platinum electrodes connected by a pH-meter (4), are dipped into the reducing (1) and the oxidizing (2) solution. Porous fritted glass (3) prevents mixing of the two solutions. The pH-meter readings correspond to Eq. (7-41), because there is no current flow (input resistance of the pH-meter is more than $10^{12}\Omega$). b—In addition, two short-circuit electrodes (5) are mounted, so that current can flow through ballast resistor R.

measuring the tension between the two platinum electrodes would quite accurately read the value calculated from Eq. (7-41) of

$$U = 0.059 \left(\lg \frac{1}{100} - \lg \frac{100}{1} \right) = -0.236 \text{ V}$$

(the factor $(RT/F) \cdot 2.303$ is 0.059 V at 25°C, 2.303 being the conversion factor to common logarithms). This arrangement is not equivalent to real soil, since the electric circuit is only closed by the very high input resistance of the pH-meter ($\sim 10^{12}$ Ω), which reduces the current flow virtually to zero. If two additional electrodes are introduced, as shown in Fig. 7-15b, connected by a resistance R, a current flows through the resistor and causes a voltage drop in R and in the solution.

If R_L is the sum of electric resistances in both solutions, the ohmic voltage drop is $\Delta U = I \cdot R + I \cdot R_L$. In measuring soil potential, R_L corresponds approximately to the resistance of the soil between the two measuring points. The voltage ΔU is opposed to the voltage U, so that (neglecting the diffusion potential) the measured voltage U_{eff} is smaller than the theoretical value U without current:

$$U_{\text{eff}} = U_{1,2} - \Delta U = U_{1,2} - I \cdot R - I \cdot R_L \qquad (7\text{-}42)$$

In a more precise treatment, the so-called overvoltages and diffusion potentials have to be deducted from $U_{1,2}$. However, for practical calculations this is of no significance, since the overvoltages and diffusion potentials in the soil are not known.

Experimentally a linear decrease of U_{eff} with decreasing R is found if $R \gg R_L$. In contrast, U_{eff} is determined by R_L; if $R_L \gg R$, then

$$\Delta U \cong I \cdot R_L \qquad (7\text{-}43)$$

and the measurable tension between the two electrodes disappears almost completely. This voltage breakdown by an outside short-circuit is quite important in connection with the stabilization of landslides by the short-circuit conductor method, because in this way the driving force of the electroosmotic water transport is eliminated (Veder 1957 etc., and 6.1.4.4).

A very important parameter for the concentrations of Fe^{3+} and Fe^{2+} in the pore water is the *in situ* pH-value of the pore water. This pH may be strongly influenced by rain water and especially by waste waters seeping in (7.2.2.3.2). The soil always contains an excess of solid and therefore electrochemically inactive iron(III) and iron(II) oxide hydrates. The solubility of these compounds and thereby the active concentrations of Fe^{3+} and Fe^{2+} in the pore water now are functions of the pH, because the compounds are precipitated by hydroxyl ions and dissolved by hydrogen ions:

$$FeO \cdot OH + 3 H^+ \rightarrow Fe^{3+} + 2 H_2O$$

$$FeO + 2 H^+ \rightarrow Fe^{2+} + H_2O$$

$$(7\text{-}43a)$$

$$Fe^{3+} + 3\,OH^- \rightarrow FeO \cdot OH + H_2O$$

$$Fe^{2+} + 2\,OH^- \rightarrow FeO + H_2O \tag{7-43b}$$

Generally, the solubility product L determines the solubility of ionogenic compounds in water. The solubility product for $FeO \cdot OH$ and $FeO \cdot H_2O$ is

$$L_{FeO \cdot OH} = c_{Fe^{3+}} \cdot c_{OH^-}^3 = 10^{-37.4} \tag{7-44a}$$

$$L_{FeO \cdot H_2O} = c_{Fe^{2+}} \cdot c_{OH^-}^2 = 10^{-16.5} \tag{7-44b}$$

Since according to 7.2.2.3.2 the relation between concentration of hydroxyl ions c_{OH^-} and concentration of hydrogen ions c_{H^+} is known, and since from c_{H^+} the pH is easily calculated, it is also possible to calculate the relation between iron(III)- and iron(II)-ion concentration and pH, if solid $FeO \cdot OH$ or $FeO \cdot H_2O$ are present:

$$c_{Fe^{3+}} = 10^{4.54} - 10^{-3pH} \tag{7-45a}$$

$$c_{Fe^{2+}} = 10^{14.56} - 10^{-2pH} \tag{7-45b}$$

Inserting (7-45a) and (7-45b) into (7-40) and simplifying, one gets the equation for the soil redox potential as a function of the *in situ* pH-value of pore water:

$$U = 0.186 - 0.059\ \text{pH}. \tag{7-46}$$

In spite of the approximations contained in (7-46), the agreement with the results of practical soil measurements is quite good, as shown in Table 7-2.

Good agreement like this is only to be expected with relatively small potential differences and in the absence of hydrogen sulfide, metal sulfides, and significant quantities of organic substances. Theory shows that the correction terms, neglected in the deductions of Eqs. (7-42), (7-43), and (7-46), increase with increasing potential differences, and consequently so do the deviations between experimental values and calculated values. If significant quantities of

Table 7-2 Experimental and calculated potentials in layered loam (sandy-clayey silt), Styria, Austria

Layer	pH of pore water	ΔU_{exp} [mV] (vs. uppermost brown layer = 0)	ΔU_{cal} [mV] (Eqn. 7-46)
Uppermost brown	9.55	0 = Reference Point	= 0
Upper blue	9.10	+ 17	+ 26
Middle brown	9.40	+ 6	+ 8
Lower blue	9.30	+ 18	+ 14
Lowermost brown	9.45	+ 13	+ 5

hydrogen sulfide H_2S, metal sulfides, and organic substances are present, Eq. (7-46) deviates from the measured values because iron(III) ions are reduced by hydrogen sulfide to iron(II) ions, and because in Eq. (7-46) the solubility of iron(II) sulfide is neglected.

Correlation between soil potentials and landslides. As outlined in 7.2.2.4, water migrates in clay soils by natural electroosmosis from zones with positive to zones with negative redox potential. Since electroosmosis produces an excess pressure, this water can be taken up by clay even under an existing consolidation pressure and cause swelling and a decrease of shear strength. Consequently, it may be expected that slip surfaces will preferably develop where the local soil potential is significantly negative. Measurements taken in slide zones confirm this assumption (Fig. 7-16). Following Eq. (7-46), the local soil potential becomes more negative with increasing pH-value and consequently the pH-value should be higher in the slip zone than in the adjacent soil. This conclusion is confirmed by experiment, as shown in Fig. 7-17. Especially in the immediate vicinity of slip surfaces, the potential gradient is high, which explains the increased water content and the proportionally decreased shear strength. The

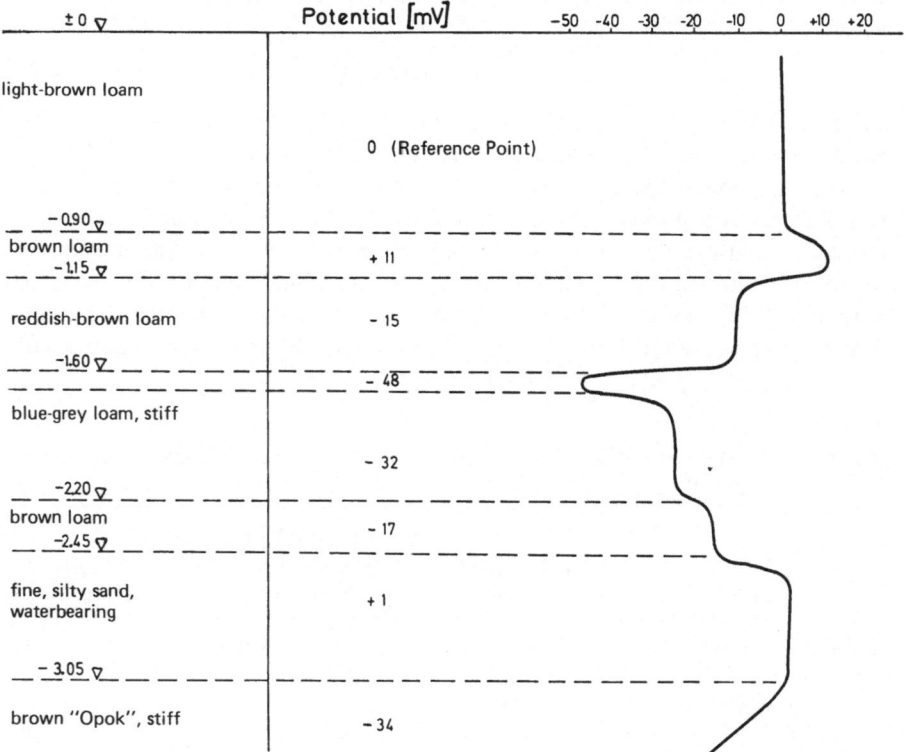

Figure 7-16 Potential measurements in a trench in a slide, taken immediately after excavation. *Opok* = overconsolidated silt with varying shares of sand and/or clay.

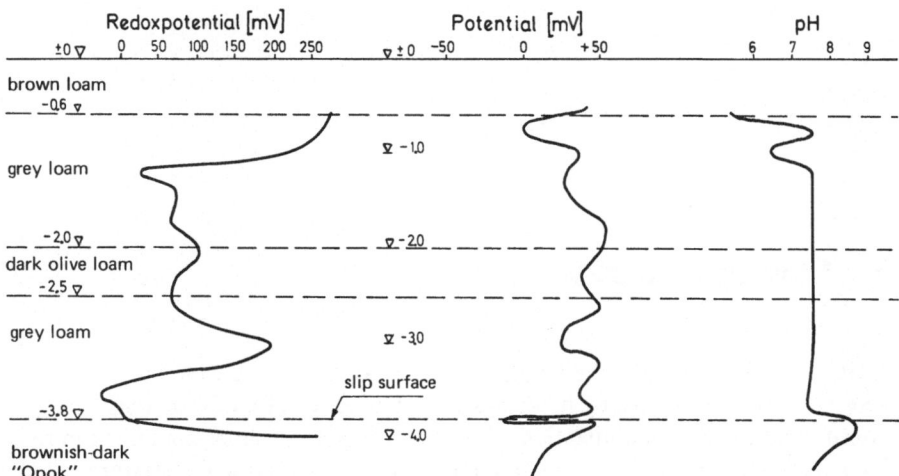

Figure 7-17 Soil profile, redox potential profiles and pH-value in a slide slope in Sarukuyoji, Japan. *Opok* = overconsolidated silt with varying shares of sand and/or clay.

effects of such potential differences increase with decreasing permeability coefficient k_h of the soil (7.2.2.4), since the pore-water pressure produced by electroosmosis is approximately inversely proportional to k_h. Soils with a high content of very finely grained clay minerals, especially montmorillonite, tend to develop this type of landslide. Owing to the mechanism shown, these slides may be stabilized quite successfully by the short-circuit conductor method.

8 Closing Remarks

Here it should be emphasized again that this book covers a wide but, owing to geographical limitations, by no means complete range of landslide types and of alternatives for their stabilization. However, the presentations contained in the book itself may easily be supplemented from the international literature listed in the references, which deals with landslide phenomena in almost all parts of the world.

In all fields of science in general, and in the field of geotechnical engineering in particular, nothing is ever complete and nothing is ever final. In this sense my book offers the geotechnical engineer the opportunity to make use of past experiences and findings, to develop them further, and to work towards new and even better methods for analysis and stabilization of landslides.

References

Aitchison, G. D. (1973). Problems of Soil Mechanics and Construction on Soft Clays and Structurally Unstable Soils. *Proc. 8th International Conference on Soil Mechanics and Foundation Engineering*, Moscow, Vol. 3:161–190.

Aitchison, G. D. (1973). General Report on Structurally Unstable Soils. *Proc. 8th International Conference on Soil Mechanics and Foundation Engineering*, Moscow, Vol. 3:161–190.

Andrei, S. (1976). Cacul de la Stabilité de Talus. *Proc. 6th European Conference on Soil Mechanics and Foundation Engineering*, Vienna, Vol. 2.2.

Andrei, S., and Athanasiu, R. (1980). Stress-Strain Relation in Slope Stability Analysis. *Proc. International Symposium on Landslides*, New Delhi, Vol. 1:109–112.

Andresen, A., and Bjerrum, L. (1967). Slides in Subaqueous Slopes in Loose Sand and Silt. In, Richards, A. F., ed., *Marine Geotechnique*. University of Illinois Press, Urbana: 221–239.

Aoyama, K., and Ogawa, S. (1980). Characteristics of Soils of Landslide Areas in Niigata Prefecture. *Proc. International Symposium on Landslides*, New Delhi, Vol. 1:129–138.

Arango, I., and Seed, H. B. (1974). Seismic Stability and Deformation of Clay Slopes. *Journal of Soil Mechanics and Foundations Division*, Vol. 100, No. GT2:139–156.

Arulanandan, K., Canclini, J., and Anandarajah (1979). Simulation of Earthquake Motions in Centrifuge. *Proc. ASCE Symposium on Centrifugal Modeling of Geotechnical Problems*.

Athanasiu, C. (1980). Non-Linear Slope Stability Analysis. *Proc. International Symposium on Landslides*, New Delhi, Vol. 1:259–262.

Austrobohr (1978). *Company Internal Report*, Graz, Austria.

Badura, G. (1976). Beton-Zyklopen für Hangsicherungen - nicht nur im Eisenbahnbau. *Tiefbau, Ingenieurbau, Strassenbau*, Vol. 18, No. 2:84–86.

Bailey, W. A., and Christian, J. T. (1969). ICES LEASE-1, A Problem Oriented Language for Slope Stability Analysis. Users Manual, Department of Civil Engineering, Massachusetts Institute of Technology, Report R69-22.

Baker, R., and Garber, M. (1977). Variational Approach to Slope Stability. *Proc. 9th International Conference on Soil Mechanics and Foundation Engineering*, Tokyo, Vol. 2:9–12.

Barden, L., McGown, A., and Collins, K. (1973). The Collapse Mechanism in Partly Saturated Soil. *Engineering Geology*, Vol. 7, No. 1:49–60.

Bauer, G. E., Deschamps, G. P., and Scott, J. D. (1980). The Effect of Shear Strength and Pore Pressure Distribution on the Stability of Natural Slopes. *International Symposium on Landslides*, New Delhi, Vol. 1:297–302.

Beasley, D. H., and James, R. G. (1975). Use of a Hopper to Simulate Embankment Construction in a Centrifugal Model. Cambridge University Engineering Department No. CUED/C-Soils/TR-22.

Bell, J. M. (1966). Dimensionless Parameters for Homogeneous Earth Slopes. *ASCE Proc.* Vol. 92, SM 5:51–65.

Bell, J. M. (1968). General Slope Stability Analysis. *Journal of the Soil Mechanics and Foundations Division,* SM 6:1253–1270.

Bell, J. M. (1969). Noncircular Sliding Surfaces. *Journal of Soil Mechanics and Foundations Division,* SM 3:829–844.

Bentz, A. (1961). *Lehrbuch angewandter Geologie,* Vol. 1. Ferdinand Enke Verlag, Stuttgart.

Bernander, S. (1978a). Progressive Failures in Normally Consolidated Clays. Seminar, Chalmers Technical University, Department of Soil Mechanics and Foundation Engineering.

Bernander, S. (1978b). Brittle Failures in Normally Consolidated Soils. *Vägoch Vattenbyggaren,* No. 8–9:49–54.

Bertini, T., and Rossi-Dorai, M. (1980). Landslides in a Structurally Complex Formation, *Proc. International Symposium on Landslides,* New Delhi, Vol. 1:199–206.

Bhandari, R. K. (1977). Some Typical Landslides in the Himalaya. *Proc. 2nd International Symposium on Landslides and Their Control,* Japan Society of Landslides, Tokyo:1–32.

Bhandari, R. K. (1980). Prediction of Landslide Behaviour: Instrumentation, etc. *Proc. International Symposium on Landslides,* New Delhi, Vol. 2:27–30.

Bhandari, R. K., and Gurdev Singh (1972). On the Formation of Rilled-Earth-Buttresses in Boulder Conglomerate Formation of Upper Siwalkis. *Bulletin of Indian Geological Association,* Vol. 5:47–49.

Bhandari, R. K., and Natarajan, T. K. (1980). A Major Landslide in Sikkim—Analysis, Correction and Efficacy of Protective Measures. *Proc. International Symposium on Landslides,* New Delhi, Vol. 1:397–402.

Bhandari, R. K., Sreenivasulu, V., and Sudhir Sharma (1980). Simple Devices for Monitoring Problematic Slopes and Buildings. *Proc. International Symposium on Landslides,* New Delhi, Vol. 1:343–348.

Biczysko, S. J., and Starewski, K. (1977). Daventry Bypass Landslip. A Lesson for Road Engineers. *Ground Engineer,* Vol. 1:23–25.

Bishop, A. W. (1954). *Proc. European Conference on Stability of Earth Slopes,* Stockholm:25–27.

Bishop, A. W. (1955). The Use of the Slip Circle in the Stability Analysis of Slopes. *Geotechnique,* Vol. 5:7–17.

Bishop, A. W. (1957). Some Factors Controlling the Pore Pressure Set Up During Construction of Earth Dams. *Proc. 4th International Conference on Soil Mechanics and Foundation Engineering,* London.

Bishop A. W. (1967). Special Problems in Slope Stability. *Journal of the Soil Mechanical Division,* SM 4:499–526.

Bishop, A. W. (1967a). Progressive Failure With Special Reference to the Mechanism Causing It. *Proc. Geotechnical Conference,* Oslo.

Bishop, A. W. (1973). The Stability of Tips and Soil Heaps. *Quarterly Journal of Engineering Geology,* Vol. 6, No. 4:335–376.

Bishop, A. W., and Henkel, D. J. (1962). *The Measurement of Soil Properties in the Triaxial Test.* Edward Arnold Publishers, London.

Bishop, A. W., and Morgenstern, N. R. (1960). Stability Coefficients for Earth Slopes. *Geotechnique,* Vol. 10, No. 4:129–150.

Bjerrum, L. (1954). Geotechnical Properties of Norwegian Marine Clay. *Geotechnique* 4:49–69.

Bjerrum, L. (1967). Progressive Failure in Slopes of Overconsolidated Plastic Clay and Clay Shales. *Journal of the Soil Mechanics and Foundations Division,* Vol. 93, SM 5:3–49.

Bjerrum, L. (1967a). Progressive Failure - With Special Reference to the Mechanism Causing It. *Proc. Geotechnical Conference,* Oslo, Vol. 2:142–150.

Bjerrum, L. (1969). Stability of Natural Slopes and Embankment Foundations. *Main Session 5, 7th International Conference on Soil Mechanics and Foundation Engineering,* Mexico, Vol. 3:411.

Bjerrum, L. (1973). Problems of Soil Mechanics and Construction on Soft Clays and Structurally Unstable Soils. *Proc. 8th International Conference on Soil Mechanics and Foundation Engineering,* Moscow, Vol. 3:111–159.

Bjerrum, L., and Grande, L. (1976). Stability of Temporary and Permanent Slopes. *General Report, Session I/1, 6th European Conference on Soil Mechanics and Foundation Engineering,* Vienna, Vol. 2.1:11–26.

Bjerrum, L., and Kenney, T. C. (1967). Effect of Structure on the Shear Behaviour of Normally Consolidated Quick Clays. *Proc. Geotechnical Conference on Shear Strength Properties of Natural Soils and Rocks,* Norwegian Geotechnical Institute, Oslo, Vol. 2:19–27.

Bjerrum, L., Løken, T., Heiberg, S., and Foster, R. (1969). A Field Study of Factors Responsible for Quick Clay Slides. *Proc. 7th International Conference on Soil Mechanics and Foundation Engineering,* Mexico City, Vol. 2:531–540.

Blight, G. (1977). Slopes and Excavations in Residual Soils. *Proc. 9th International Conference on Soil Mechanics and Foundation Engineering,* Tokyo, Vol. 2.

Blong, R. J. (1973). Relationships Between Morphometric Attributes of Landslides. *Zeitschrift für Geomorphologie,* Supp. Vol. 18:66–77.

Blong, R. J. (1973a). A Numerical Classification of Selected Landslides of the Debris Slide-Avalanche-Flow Type. *Engineering Geology,* Vol. 7, No. 2:99–114.

Booker, J. K., and Davis, E. H. (1972). A Note on a Plastic Solution to the Stability of Slopes in Homogeneous Clays. *Geotechnique,* Vol. 22, No. 4:509–513.

Borowicka, H., sen. (1957). Über die Standsicherheit von Böschungen. *Österreichische Ingenieur-Zeitschrift,* Österreichischer Ingenieur- und Architekten Verein, Wien.

Borowicka, H., sen. (1959). Über die Sicherheit im Grundbau. Department of Foundation Engineering and Soil Mechanics, Vienna Technical University, No. 2:36–57.

Borowicka, H., sen. (1963). Der Wiener Routinescherversuch. Department of Foundation Engineering and Soil Mechanics, Vienna Technical University, Report No. 5.

Borowicka, H., sen. (1965). Die Standsicherheit einer Böschung in Theorie und Praxis. *Der Bauingenieur, Zeitschrift für das gesamte Bauwesen,* Nr. 1:21–26.

Borowicka, H., sen. (1968). Ein einfaches und widerspruchsfreies Verfahren zur Ermittlung der Standsicherheit von Böschungen. Department of Foundation Engineering and Soil Mechanics, Vienna Technical University, Report No. 10.

Borowicka, H., sen. (1970). Ein statisch einwandfreies Verfahren zur Ermittlung der Standsicherheit einer Böschung. *Der Bauingenieur, Zeitschrift für das gesamte Bauwesen,* No. 9:307–313.

Borowicka, H. R., jun. (1968). Die Rutschungen an der Autobahn Salzburg–Wien, Thesis, Vienna University.

Brandl, H. (1973). Stabilization of Slippage-Prone Slopes by Lime Injections. *8th International Conference on Soil Mechanics and Foundation Engineering,* Moscow, Vol. 4.3:300–301.

Brandl, H. (1976a). Böschungssicherung und Sanierung von Rutschungen. *Straße und Autobahn,* Vol. 27, No. 5:179–204, and No. 6:234–240.

Brandl, H. (1976b). Die Sicherung von hohen Abschnitten in rutschgefährdeten Verwitterungsböden. *Proc. 6th European Conference on Soil Mechanics and Foundation Engineering,* Vol. 1.1:19–28.

Brinch Hansen, J., and Lundgren, H. (1960). *Hauptprobleme der Bodenmechanik.* Springer, Berlin-Göttingen-Heidelberg.

Broms, B. B. (1969). Stability of Natural Slopes and Embankment Foundations (Göta River Valley). *Proc. 7th International Conference on Soil Mechanics and Foundation Engineering,* Vol. 3:385–394.

Broms, B. B. (1975). *Landslides.* Foundation Engineering Handbook, Van Nostrand-Reinhold, New York: 373–401.

Broms, B. B. (1978). Translationary Slips in Soft Clay (Slides in Göta River Valley). Unpublished.

Broms, B. B. (1979). Problems and Solutions to Construction in Soft Clays. *Proc. 6th Asian Regional Conference on Soil Mechanics and Foundation Engineering,* Singapore, Vol. 2:35–38.

Broms, B. B., and Bennermark, H. (1967). Discussion. *Proc. Geotechnical Conference,* Oslo, Vol. 2:118–120.

Broms, B. B., and Boman, P. (1976) Stabilization of Deep Cuts With Lime Columns. *Proc. 6th European Conference on Soil Mechanics and Foundation Engineering,* Vienna, Vol. 1.1:207–210.

Broms, B. B., and Stal, T. (1980). Landslides in Sensitive Clays. *Proc. International Symposium on Landslides,* New Delhi, Vol. 2:39–66.

Cabrera, J. G., and Smalley, I. J. (1973). Quick Clays as Products of Glacial Action: A New Approach to their Nature, Geology, Distribution, and Geotechnical Properties. *Engineering Geology,* Vol. 7, No. 2:115–133.

Caquot, A., and Kerisel, J. (1956). *Traité de Mécanique des Sols.* Gauthier-Villars, Paris.

Carter, R. M. (1975). A Discussion and Classification of Sub-Aqueous Mass Transport with Particular Application to Grain-Flow, Slurry-Flow, and Fluxoturbidities. *Earth-Science Review,* Vol. 11, No. 2:145–177.

Carter, R. K., Lovell, C. W. Jr., and Harr, M. E. (1971). Computer.Oriented Stability Analysis of Reservoir Slopes. Water Resources Research Center, Purdue University, Technical Report No. 17:1–28.

Carson, M. A. (1977). On the Retrogression of Landslides in Sensitive Muddy Sediments. *Canadian Technical Journal,* Vol. 14, No. 4:582–602.

Carson, M. A. (1979). On the Retrogression of Landslides in Sensitive Muddy Sediments: Reply. *Canadian Geotechnical Journal,* Vol. 16, No. 2.

Casagrande, A. (1976). Liquefaction and Cyclic Deformation of Sands. *Harvard Soil Mechanics Series,* No. 88.

Casagrande, A., and Shannon, W. L. (1947/48). Research on Stress Deformation and Strength Characteristics of Soils and Soft Rocks Under Transient Loading. *Harvard Soil Mechanics Series,* No. 31.

Casagrande, A., Shannon, W. L. and Matich, M.A.J. (1961). Electro-Osmotic Stabilization of a High Slope in Loose Saturated Silt. *Proc. 5th International Conference on Soil Mechanics and Foundation Engineering, Paris*, Vol. 2:555–561.

Casagrande, L. (1948). Electroosmosis. *Proc. 2nd International Conference on Soil Mechanics and Foundation Engineering, Rotterdam*, Vol. 1:218–223.

Casagrande, L. (1949). Electroosmosis in Soils. *Geotechnique* 1, No. 3:159–177.

Casagrande, L. (1962). Electroosmosis and Related Phenomena. *Revista Ingeneria Mexico*, Vol. 31, No. 2:1–62.

Castillo, E., and Revilla, J. (1977). The Calculus of Variations and the Stability of Slopes. *Proc. 9th Conference of Soil Mechanics and Foundation Engineering, Tokyo* 2:25–30.

Castro, G. (1975). Liquefaction and Cyclic Mobility of Saturated Sands. *Journal of Geotechnical Engineering Division*, Vol. 101, No. GT6:551–569.

Chandler, R. J. (1972). Lias Clay: Weathering Processes and Their Effect on Shear Strength. *Geotechnique*, Vol. 22:403–431.

Chandler, R. J. (1974). Lias Clay: The Long-Term Stability of Cutting Slopes. *Geotechnique*, Vol. 24, No. 1:21–38.

Chandler, R. J., and Skempton, A. W. (1974). The Design of Permanent Cutting Slopes in Stiff Fissured Clays. *Geotechnique*, Vol. 24, No. 4:457–466.

Chatterjee, B. (1975). Role of Geological Structures in the Landslide Problems With Reference to the North-Eastern Himalayas. *Proc. Seminar on Landslides and Toe Erosion Problems With Special Reference to Himalayan Regions*, Gangtok:89

Chen, W., and Snitbhan, N. (1975). On Slip Surface and Slope Stability Analysis. *Soils and Foundations*, Vol. 15, No. 3:41–49.

Cheney, J. A., and Fragaszy, R. J. (1980). Short-Term Stability of Over-Consolidated Clay Slopes. *International Symposium on Landslides*, New Delhi, Vol. 1:149–156.

Chowdhury, R. N. (1978a). Slope Analysis. *Developments in Geotechnical Engineering*, Vol. 22, Elsevier, New York.

Chowdhury, R. N. (1978b). Factors Affecting Slope Performance. *Geotechnical Engineering and Environmental Control*, Vol. 2.

Chowdhury, R. N. (1980). Recent Progress in Evaluation and Control of Landslides. *International Symposium on Landslides*, New Delhi, Vol. 1:313–320.

Cleaves, A. B. (1961). *Landslides Investigations: A Field Handbook for Use in Highway Locations and Design*. U.S. Bureau of Public Roads.

Clough, R. W., and Woodward, R. J. (1967). Analyses of Embankment-Stresses and Deformations. Journal of Soil Mechanics and Foundations Division, *Proc. of the American Society of Civil Engineers*, New York, SM 4:529–549.

Coatsworth, A. M., and Tabb, R. (1978). Discussion of "Usage, Requirements, and Features of Slope Stability Computer Software, (Canada, 1977)." *Canadian Geotechnical Journal*, Vol. 15:435–436.

Cole, K. W., and Burland, J. B. (1972). Observations of Retaining Wall Movements Associated With a Large Excavation. *Proc. 5th European Conference on Soil Mechanics and Foundation Engineering*, Madrid: 327–342.

Colleselli, F. (1977). Consideranzioni sui metodi di verifica della stabilità dei pendi e sulla determinazione del coefficiente di sicurezza. *Memorie e studi dell'istituto di construzioni marittime e di geotecnica* No. 128, Università di Padova, Facultà di Ingegneria.

Collins, T. (1973). Bibliography of Recent Publications on Slope Stability, Landslide. *The Slope Stability Review*, Vol. 1, No. 1: 28–37.

Colwell, R. N. (1980). Manual of Photographic Interpretation. American Society of Photogrammetry, Falls Church, Va.

Costa Nunes, A. J. da (1969). Landslides of Decomposed Rock Due to Intense Rainstorms. *Proc. 7th International Conference on Soil Mechanics and Foundation Engineering*, Mexico, Vol 2:547–554.

Cottechia, V. (1977). Systematic Reconnaissance, Mapping and Registration of Slope Movements. *International Symposium on Landslides and Other Mass Movements*, Prague, General Report, Theme 1, IAEG Bull. Nr. 17:5–37.

Cottechia, V., and Melidoro, G. (1974). Some Principal Geological Aspects of the Landslides of Southern Italy. *Bull. International Association of Engineering Geology* No. 9:23–32.

Coulter, H. W., and Migliaccio, R. R. (1966). Effects of the Earthquake of March 27, 1964 at Valdez, Alaska. *U.S. Geological Survey*, Professional Paper 542-C.

Crawford, C. B. (1963). Cohesion in an Undisturbed Sensitive Clay. *Geotechnique*, Vol. 13, No. 2:132–146.

Crawford, C. B. (1968). Quick Clays of Eastern Canada. *Engineering Geology*, Vol. 2, No. 4:239–265.

Crawford, C. B., and Eden, W. J. (1963). *National Research Council Canada*, Engineering Geology Case Histories, No. 4:45–50.

Crawford, C. B., and Eden, W. J. (1967). Stability of Natural Slopes in Sensitive Clay. *Proc. ASCE, Journal of Soil Mechanics and Foundation Engineering Division*, Vol. 93, No. SM4:419–436.

Crozier, M. J. (1969). Earthflow Occurrence During High Intensity Rainfall in Eastern Otago (New Zealand). *Engineering Geology*, Vol. 3, No. 4:325–334.

Crozier, M. J. (1972). Some Problems in the Correlation of Landslide Movement and Climate. *International Geography, Proc. 22nd International Geographical Congress*, Montreal, University of Toronto Press, Ontario, Vol. 1:90–93.

Crozier, M. J. (1973). Techniques for the Morphometric Analysis of Landslips. *Zeitschrift für Geomorphologie*, Vol. 1, No. 1:78–101.

Daido, A. On the Occurrences of Mud-Debris Flow. *Bulletin of the Disaster Prevention Research Institute*, Kyoto University, Japan, Vol. 21, Part 2, No. 187:109–135.

D'Appolonia, E., Alperstein, R., and D'Appolonia, D. J. (1967). Behaviour of a Colluvial Slope. *Journal of Soil Mechanics and Foundations Division*, Vol. 93, No. SM4:447–473.

Dascal, O., and Tournier, J. P. (1975). Embankments on Soft and Sensitive Clay Foundation. *Proc. ASCE, Journal of the Geotechnical Engineering Division*, Vol. 101, No. GT3:297–314.

Dascal, O., Tournier, J. P., Tavenas, F., and La Rochelle, P. (1972). Failure of a Test Embankment on Sensitive Clay. *Proc. of the Specialty Conference on Performance of Earth and Earth-Supported Structures*, ASCE, Vol. 1, Part 1:129–158.

Datye, K. R. (1980). Landslide Control Measures and Their Efficacy: Case Histories, Etc. *Proc. International Symposium on Landslides*, New Delhi, Vol. 2:31–35.

Datye, K. R., and Nagaraju, S. S. (1980). Flexible Piles for Landslides Control. *Proc. International Symposium on Landslides*, New Delhi, Vol. 1:439–442.

Deere, D. U., and Patton, D. F. (1971). Slope Stability in Residual Soils. *Proc. 4th Pan-American Conference on Soil Mechanics and Foundation Engineering*, San Juan, Vol. 1:87–170.

Deguchi, M., Nagase, M., and Nansaki, N. (1980). Downshift of Sliding Surface in Cutting Slope With Restriction Piles. *Proc. International Symposium on Landslides*, New Delhi, Vol. 1:443–448.

D'Elia, B. (1975). Aspetti Meccanici delle Frane tipo "Colata". *Rivista Italiana di Geotecnica,* 9:32–42.

D'Elia, B. (1977). Geotechnical Complexity of Some Italian Variegated Clay Shales. *International Symposium on Geotechnics of Structurally Complex Formations,* Panel Discussion, Session II, Capri, Vol. 2.

D'Elia, B. (1979). Caratteri Cinematici delle Colate: Interventi di Stabilizzazione. *Rivista Italiana di Geotecnica.*

D'Elia, B., Esu, F., Grisolia, M., and Tancredi, G. (1980). Stability of Slopes in Clay Shale Formation. *Proc. International Symposium on Landslides,* New Delhi, Vol. 1:157–164.

Denness, B. (1972). Reservoir Principle of Mass Movement. *Institute of Geological Sciences, UK,* Rep. 72/7.

Deschamps, G. P. (1979). The Effect of Shear Strength and Pore Pressure Distribution on the Stability of Slopes, M.A.Sc. Thesis, University of Ottawa, Ottawa.

Dhakharia, K. D. (1980). Value of Photogrammetry for Landslide Studies. *Proc. International Symposium on Landslides,* New Delhi, Vol. 1:95–98.

Dishaw, H. E. (1967). Massive Landslides. *Photogrammetric Engineering,* Vol. 32, No. 6:603–609.

Dube, A. K., Jethwa, J. L., and Singh, B. (1980). Monitoring of Slope Movements Around Power House Pit by Borehole Extensometers. *Proc. International Symposium of Landslides,* New Delhi, Vol. 1:361–364.

Duncan, J. M. (1972). Finite Element Analysis of Stresses and Movements in Dams, Excavations and Slopes. Application of the Finite Element Method in Geotechnical Engineering: A Symposium, Descai, C. S., ed. U.S. Army Engineering Waterways Experiment Station, Vicksburg, Miss.: 267–326.

Duncan, J. M., and Seed, H. B. (1966). Strength Variation Along Failure Surfaces in Clay. *Journal of Soil Mechanics and Foundations Division,* Vol. 2, No. SM6:81–104.

Duncan, J. M., and Wright, S. G. (1980). The Accuracy of Equilibrium Methods of Slope Stability Analysis. *Proc. International Symposium on Landslides,* New Delhi, Vol. 1:247–254.

Dunlop, P. and Duncan, J. M. (1970). Development of Failure Around Excavated Slopes. *Journal of Soil Mechanics and Foundations Division,* Vol. 96, No. SM2:471–493.

Dunlop, P., Duncan, J. M. and Seed, H. B. (1968). Finite Element Analysis of Slopes in Soil. Department of Civil Engineering, University of California, Berkeley, Report TE-68-3.

Dusseault, M. B., and Morgenstern, N. R. (1979). Locked Sands. *Quarterly Journal of Engineering Geology,* Vol. 12:117–131.

Dylik, J. (1967). Solifluxion, Congelifluxion, and Related Slope Processes, *Geografisky Annaler,* Vol. 49A, No. 2-4:167–177.

Dysli, M. (1977). Swiss Use of the Computer in Soil Mechanics. *Proc. of the Specialty Session on Computers in Soil Mechanics: Present and Future; 9th International Conference on Soil Mechanics and Foundation Engineering,* Tokyo:268–288.

Eden, W. J. (1971). *Landslides in Clays.* National Research Council of Canada, Canadian Building Digest, CBD 143.

Eden, W. J. (1975). Mechanism of Landslides in Leda Clay With Special Reference to the Ottawa Area. *Proc. 4th Guelph Symposium on Geomorphology,* Guelph, Ontario:159–171.

Eden, W. J. (1977). Evidence of Creep in Steep Natural Slopes of Champlain Sea Clay. *Canadian Geotechnical Journal,* Vol. 14, No. 4:620–627.

Eden, W. J., and Crawford, C. B. (1957). Geotechnical Properties of Leda Clay in the Ottawa Area. *Proc. 4th International Conference on Soil Mechanics,* London, Vol. 1:22–27.

Eden, W. J., and Jarret, P. M. (1971). Landslide at Orleans, Ontario. National Research Council of Canada, Technical Paper No. 321.

Eden, W. J., and Kubota, J. K. (1961). Some Observations on the Measurement of Sensitivity of Clays. *Proc. ASTM,* Vol. 61:1239–1249.

Eden, W. J., and Mitchell, R. J., (1970). The Mechanics of Landslides in Leda Clay. *Canadian Geotechnical Journal,* Vol. 7, No. 3:285–296.

Eden, W. J. and Mitchell, R. J. (1973). Landslides in Sensitive Marine Clay in Eastern Canada. *Highway Research Record,* No. 463:18–27.

Eide, O., and Holmberg, S. (1972). Test Fills to Failure on the Soft Bangkok Clay. *Proc. ASCE Specialty Conference on Performance of Earth and Earth-Supported Structures,* Purdue University, Lafayette, Ind., Vol. 1:158–180. (*also published by Norwegian Geotechnical Institute,* Publication No. 95).

Eigenberger, K. (1972). Standsicherheit von Böschungen. Thesis, Department of Soil Mechanics, Rock Mechanics and Geotechnical Engineering, Graz Technical University, No. 2, parts I and II.

Enderli, M., Llorca, I., and Muzás, F. (1977). Landslide Stabilization by Means of Anchorages. *Proc. 9th International Conference on Soil Mechanics and Foundation Engineering,* Tokyo, Vol. 2:59–62.

Endicaott, L. J. (1971). A Centrifugal Model Test of the Trial Embankment at King's Lynn. Cambridge University Engineering Department, Report No. CUED/C-Soils/TR-12.

Escario, V. (1955) Errors Arising from the Simplified Method of Slices. Discussion 1 and 2. *Geotechnique*:48–54/86–87.

Escario, V. (1961). Errors Arising from the Simplified Method of Slices. *Proc. 5th International Conference on Soil Mechanics and Foundation Engineering,* Vol. 2:585–591.

Esu, F. (1966). Short-Term Stability of Slopes in Unweathered Jointed Clays. *Geotechnique,* Vol. 16, No. 4:321–328.

Esu, F. (1977). Behaviour of Slopes in Structurally Complex Formations. *International Symposium on Geotechnics of Structurally Complex Formations,* General Report, Session IV, Capri, Italy, Vol. II.

Esu, F., and Calabresi, G. (1969). Slope Stability in an Overconsolidated Clay. *Proc. 7th International Conference on Soil Mechanics and Foundations Engineering,* Mexico City, Vol. 2:555–563.

Esu, F., and D'Elia (1976). Time-Dependent Behaviour of Deep Excavations in Overconsolidated Jointed Clays. *Proc. 6th European Conference on Soil Mechanics and Foundation Engineering,* Vienna, Vol. 1:41–44.

Feld, J. (1971). Discussion of Slope Stability in Residual Soils (Session 2). *Proc. 4th Pan-American Conference on Soil Mechanics and Foundation Engineering,* San Juan; American Society of Civil Engineers, New York, Vol. 3:125.

Fellenius, B. (1955). The landslide at Guntorp. *Geotechnique,* Vol. 5, No. 1:120–125.

Fellenius, W. (1926/1940). *Erdstatische Berechnungen mit Reibung und Köhasion (Adhäsion) und unter Annahme kreiszylindrischer Gleitflächen.* Verlag Wilhelm Ernst & Sohn, Berlin.

Fellner, L. (1958). Beitrag zur Berechnung des Sicherheitsgrades gegen Rutschung von Erdkörpern. *Der Bauingenieur, Zeitschrift für das gesamte Bauwesen*, No. 3:105-108.

Fellner, L. (1959). Die kritische Rutschsicherheit. *Der Bauingenieur, Zeitschrift für das gesamte Bauwesen*, Springer-Verlag Berlin, No. 9:353-357.

Fisher, C. P., Leith, C. J., and Deal, D. C. (1965). An Annotated Bibliography on Slope Stability and Related Phenomena. North Carolina State Highway Commission and U. S. Bureau of Public Roads, NTIS, Springfield, Va.

Fleming, R. W., Spencer, G. S., and Banks, D. C. (1970). Empirical Study of Behavior of Clay Shale Slopes. U.S. Army Engineer Nuclear Cratering Group, Technical Report 15, Vols. 1 and 2.

Fletscher, G.F.A. (1965). Standard Penetration Test: Its Uses and Abuses. *Journal of Soil Mechanics and Foundations Division*, Vol. 91, No. SM4:67-75.

Forrester, K., and Nyland, G. (1980). Two Landslides on New South Wales Highways. *Proc. International Symposium on Landslides*, New Delhi, Vol. 1:181-184.

Franke, E. (1967). Einige Bemerkungen zur Definition der Standsicherheit von Böschungen und der Geländebruchsicherheit beim Lamellenverfahren. *Die Bautechnik, Fachzeitschrift für das gesamte Bauingenieurwesen*, No. 12:415-419.

Fredlund, D. G. (1974). Slope Stability Analysis. *User's Manual*, Computer Documentation CE-4, Transportation and Geotechnical Group, Department of Civil Engineering, University of Saskatchewan, Canada, December.

Fredlund, D. G. (1978). Usage, Requirements and Features of Slope Stability Computer Software (Canada 1977). *Canadian Geotechnical Journal*, Vol. 15:83-95.

Fredlund, D. G. (1980). Use of Computers for Slope Stability Analysis. *Proc. International Symposium on Landslides*, New Delhi, Vol. 2:129-138.

Fredlund, D. G., and Krahn, J. (1977). Comparison of Slope Stability Methods of Analysis. *Canadian Geotechnical Journal*, Vol. 14, No.3:429-439.

de Freitas, M. H., and Watters, R. J. (1973). Some Field Examples of Toppling Failure. *Geotechnique*, Vol. 23, No.4:495-514.

Fritsch, E., and Prodinger, W. (1980). Stabilization of Landslides by Means of Permanently Effective Drains. *Proc. International Symposium on Landslides*, New Delhi, Vol. 1:383-388.

Fritsch, V. (1960). *Geoelektrische Baugrunduntersuchungen*. VEB Verlag für Bautechnik, Berlin.

Fritsch, V. (1976). Chemische Bodenverfestigung mit Hilfe der Elektrokinese. *Österreichische Ingenieurzeitung*, Vol. 19, No. 1:1-7.

Fröhlich, O. K. (1949). Über den Sicherheitsgrad von Böschungen, Dämmen und seitlich gestützten Erdkörpern gegen Rutschungen auf kreiszylindrischer Gleitfläche. *Österreichische Bauzeitschrift*, No. 5:69-74.

Fröhlich, O. K. (1954). General Theory on Stability of Slopes. *Proc. European Conference on Stability of Earth Slopes*, Stockholm, Vol. 3:21-25.

Fröhlich, O. K. (1955). General Theory on Stability of Slopes. *Geotechnique*:37-47.

Fröhlich, O. K. (1955a). Kritik der gebräuchlichsten Verfahren zur Berechnung der Sicherheit von Böschungen gegen Rutschung. *Österreichisches Ingenieur Archiv*:106-118.

Fröhlich, O. K. (1959). Anwendung von Palisadenwänden zur Übertragung von Seitenschüben auf den Untergrund. Mitteilung des Instituts für Grundbau und Bodenmechanik der Technischen Universität Wien, No. 2:3-8.

Fröhlich, O. K. (1963). Grundzüge einer Statik der Erdböschungen. *Der Bauingenieur, Zeitschrift für das gesamte Bauwesen*, No. 10:371-378.

Fujita, T. (1980). Slope Analysis of Landslides in Shikoku, Japan. *Proc. International Symposium on Landslides,* New Delhi, Vol. 1:169–176.

Fukuoka, M. (1980). Instrumentation: Its Role in Landslide Prediction and Control. *Proc. International Symposium on Landslides,* New Delhi, Vol. 2:139–153.

Fukuoka, M., Yoshida, Y., and Masuda, T. (1977). Residual Strength and Frictional Resistance of Sliding Soil Masses. *Proc. 9th International Symposium on Landslides Control,* Tokyo:11–17.

Gagnon, H. (1975). Remote Sensing of Landslide Hazards on Quick Clays of Eastern Canada. *Proc. 10th International Symposium on Remote Sensing of Environment,* Environmental Research Institute of Michigan, Ann Arbor :803–810.

Gaur, V. K. (1980). Geological Aspects and Seismological Relations of Landslides and Other Types of Mass Movements. Proc. International Symposium on Landslides, New Delhi, Vol. 2:3–5.

Gedney, D. S., and Weber, Jr., W. G. (1978). Design and Construction of Soil Slopes. *Landslides: Analysis and Control,* National Academy of Sciences, Washington, D.C., Special Report 176:172–191.

Ghosh, D. K. (1980). Geological Influence of Landslides and Other Mass Movements in the Planning of Baira-Baledh-Siul Link Project, Himachal Pradesh, India. *Proc. International Symposium on Landslides,* New Delhi, Vol. 1:19–24.

Gottstein, E. (1936). Two Examples Concerning Underground Sliding Caused by Construction of Embankments and Static Investigations on the Effectiveness of Measures Provided to Assure Their Stability. *1st International Conference on Soil Mechanics and Foundation Engineering,* Cambridge, Vol. 3, G 17:122–128.

Grabowski, Z., and Wolski, W. (1977). Efficiency of the Structure Supporting Tresna Slip. *Proc. 9th International Conference on Soil Mechanics and Foundation Engineering,* Tokyo, Vol. 2:79–82.

Gray, D. H. (1970). Effects of Forest Clear-Cutting on the Stability of Natural Slopes. *Bulletin, Association of Engineering Geologists,* Vol. 7, No. 1-2:45–66.

Gray, R. E., Gardner, G. D., and Wimberly, P. M. (1980). Stabilization of a Colluvial Slope at an Urban Site. *Proc. International Symposium on Landslides,* New Delhi, Vol. 1:389–392.

Green, G. E., and Bishop, A. W. (1969). A Note on the Drained Strength of Sand under Generalized Strain Conditions. *Geotechnique,* Vol. 19, No. 1:144–149.

Gregersen, O. (1979). Kvikkleireskredet i Rissa [The Landslide in Quick Clay at Rissa]. *Proc. Nordic Geotechnical Meeting in Helsinki* :535–548.

Gregersen, O., and Løken, T. (1975). The Baastad Quick Clay Landslide 1974. Description and Back-Calculations. Norwegian Geotechnical Institute, Internal Report.

Griffin, F. (1972). Power House Slope Stabilized by Electroosmosis. *Heavy Construction News,* Wellpoint Corp. of New York.

Gudehus, G., Kolymbas, D., and Leinenkugel, H. J. (1976). Zeitverhalten von Böschungen and Einschnitten in weichem und steifem Ton. *Proc. 6th European Conference on Soil Mechanics and Foundation Engineering,* Vienna, Vol. 1.1:51.

Gupta, S. K., and Bhandari, R. C. (1980). The Role of Aerial Photo-Interpretation for Landslides Studies in Parts of Himachal Pradesh, India. *Proc. International Symposium on Landslides,* New Delhi, Vol 1:99–104.

Hamdi Peynircioglu, A. (1969). Investigation of Landslides on a Natural Slope and Recommended Measures. *Proc. 7th International Conference on Soil Mechanics and Foundation Engineering,* Mexico, Vol. 2:645–651.

Handy, R. L., and Williams, W. W. (1967). Chemical Stabilization of an Active Landslide. *Civil Engineering,* Vol. 37, No. 8:62–65.

Hansbo, S. A. (1957). A New Approach to the Determination of the Shear Strength of Clay by the Fall-Cone Test. *Proc. Royal Swedish Geotechnical Institute,* No. 14:47.

Hansen, J. B. (1967). The Philosophy of Foundation Design: Design Criteria, Safety Factors, and Settlement Limits. *Proc. Symposium on Bearing Capacity and Settlement of Foundations,* Duke University, Durham, N.C. :9–14.

Hansen, J. B., and Gibson, R. E. (1949). Undrained Shear Strengths of Anisotropically Consolidated Clays. *Geotechnique,* Vol. 1, No. 3:189–204.

Hansen, W. R. (1965). Effects of the Earthquake of March 27, 1964, at Anchorage, Alaska. *U.S. Geological Survey,* Professional Paper 542-A:68.

Harr, M. E. (1962). *Groundwater and Seepage.* McGraw-Hill, New York.

Harr, M. E. (1977). *Mechanics of Particulate Materials: A Probability Approach.* McGraw-Hill, New York.

Hasselquist, S. J., and Schiffmann, R. L. (1975). A Computer Program For Slope Stability: New York State and Simplified Bishop Method. SNOB-1. Version 3-A, User's Manual, GESA Report D-75-18, University of Colorado Computing Center, Boulder, Colorado.

Heim, A. (1932). *Bergsturz und Menschenleben.* Verlag Fretz und Wasmut, Zürich.

Henkel, D. J. (1957). Investigations of Two Long-Term Failures in London Clay Slopes at Wood Green and Northolt. *Proc. 4th International Conference on Soil Mechanics and Foundation Engineering, London,* Vol. 2:315–320.

Henkel, D. J. (1967). Local Geology and the Stability of Natural Slopes. *Journal of Soil Mechanics and Foundation Engineering,* Vol. 92, SM4:437–446.

Henkel, D. J., and Skempton, A. W. (1955). A Landslide at Jackfield, Shropshire, in a Heavily Overconsolidated Clay. *Geotechnique,* Vol. 5, No. 2:131–137.

Hill, R. A. (1934). Clay Stratum Dried Out to Prevent Landslips. *Civil Engineering,* Vol. 4, No. 8:403–407.

Hirao, K., and Okubo, S. (1971). Studies on the Occurrence and Expansion of Landslides Caused by Ebino-Yoshimatsu Earthquake. Report of Cooperative Research for Disaster Prevention, Japan Science and Technology Agency, No. 26:157–189 (in Japanese with English summary).

Hird, C. C. (1974). Centrifugal Model Tests of Flood Embankments. Ph.D. Thesis, University of Manchester, Department of Science and Technology, U.K.

Höller, H., Homan, O., Kolmer, H., and Wirsching, U. (1977). Natürliche Gesteinsveränderungen—Probleme beim Straßen-und Tunnelbau. *Rock Mechanics,* 10:73–80.

Holm, O. S. (1961). Stabilization of a Quick-Clay Slope by Vertical Sand Drains. *Proc. 5th International Conference on Soil Mechanics and Foundation Engineering,* Vol. 2:625–627.

Holtz, R. D., and Bomann, P. (1974). A New Technique for Reduction of Excess Pore Pressures During Pile Driving. *Canadian Geotechnical Journal,* Vol. 11:423–430.

Holtz, R. D., and Massarsch, K. R. (1976). Improvement of the Stability of an Embankment by Piling and Reinforced Earth. *Proc. 6th European Conference on Soil Mechanics and Foundation Engineering,* Vol. 1-2:473–478.

Holtz, W. G. (1973). Bibliography on Landslides and Mudslides. Building Research Advisory Board, National Academy of Sciences, Washington, D.C.

Homann, O. (1975). Einige Anwendungsmöglichkeiten von Bodenstabilisierungen im steirischen Straßenbau. Special Printing from a Report of the Abteilung Geologie, Paläontologie und Bergbau, Landesmuseum Joanneum, Graz, Austria.

Horn, J. A. (1960). Computer Analysis of Slope Stability. *Journal of Soil Mechanics and Foundation Engineering,* Vol. 86, No. SM3:1–17.

Hunter, J. H., and Schuster, R. L. (1968). Stability of Simple Cuttings in Normally Consolidated Clays. *Geotechnique,* Vol. 18, No. 3:372–378.

Hurtubise, J. E., Gadd, N. R., and Meyerhof, G. G. (1957). Les eboulements de terrain dans l'est du Canada. *Proc. 4th International Conference on Soil Mechanics and Foundation Engineering.* Vol. 2:325–329.

Hutchison, J. N. (1961). A Landslide on a Thin Layer of Quick Clay at Furre, Central Norway. *Geotechnique,* Vol. 11, No. 2:69–94.

Hutchison, J. N. (1967). The Free Degradation of London Clay Cliffs. *Proc. Geotechnical Conference on Shear Strength Properties of Natural Soils and Rocks,* Norwegian Geotechnical Institute, Oslo, Vol. 1:113–118.

Hutchison, J. N. (1969). Reconsideration of the Coastal Landslides at Folkestone Warren, Kent. *Geotechnique,* Vol. 9, No. 1:6–38.

Hutchison, J. N. (1970). A Coastal Mudflow on the London Clay Cliffs at Beltinge, North Kent. *Geotechnique,* Vol. 20, No. 4:412–438.

Hutchison, J. N. (1974). Periglacial Solifluxion: An Appropriate Mechanism for Clayey Soils. *Geotechnique,* Vol. 24, No. 3:438–443.

Hutchison, J. N. (1977). Assessment of the Effectiveness of Corrective Measures in Relation to Geological Conditions and Types of Slope Movement. *Symposium on Landslides and Other Mass Movements,* Prag.

Hutchison, J. N. (1977). General Report, Theme 3. *Proc. Symposium on Landslides and Other Mass Movements,* Prag, Bull. IAEG 16:131–155.

Hutchison, J. N., and Bhandari, K. K. (1971). Undrained Loading: A Fundamental Mechanism of Mudflows and Other Mass Movements. *Geotechnique,* Vol. 21, No. 4:353–358.

Hvorlev, M. J. (1949). Subsurface Exploration and Sampling of Soils for Civil Engineering Purposes. U.S. Army Engineer Waterways Experiemnt Station, Vicksburg, Miss.

ICOS (1958). On the Elimination of Self-Potentials by Short-Circuit Conductors. *German Patent BRD 1059846* (Christian Veder—Inventor); *Italian Patent 589885* (Christian Veder—Inventor), ICOS, Milan.

ICOS (1968). The ICOS Company in the Underground Works. ICOS, Milan, Internal Report 3(a):256–258; (b):354–356.

Imai, K., Nakao, K., Aoto, H., Iihoshi, S., Kaneko, S., and Suzuki, A. (1977). Report on the Use of Steelbars in Sarukyoji Landslide. Internal Report, Taisei Company, Tokyo.

Jaky, J. (1936). Stability of Earth Slopes. *Proc. 1st International Conference on Soil Mechanics and Foundation Engineering, Harvard University, Cambridge, Mass.,* Vol. 2:200–207.

Janbu, N. (1973). *Slope Stability Computations.* The Embankment Dam Engineering, Casagrande Volume. John Wiley & Sons, New York:47–86.

Janbu, J. (1977a). State-of-the-Art Report: Slopes and Excavations. *Proc. 9th International Conference on Soil Mechanics and Foundation Engineering,* Tokyo, Vol. 2:549–566.

Janbu, N. (1977b). Failure Mechanism in Quick Clays. *Proc. Nordic Geotechnical Meeting,* Helsingfors, Finland :476–485.

Janbu, N. (1979). Mechanism of Failure in Natural and Artificial Soil Structures. *Proc. International Symposium,* Oaxaca, Mexico, Vol. 1:95–124.

Janbu, N. (1980). Critical Evaluation of the Approaches to Stability Analysis of Landslides and Other Mass Movements. *Proc. International Symposium on Landslides,* New Delhi, Vol. 2:109–128.

Janbu, N., Kjekstad, O., and Senneset, K. (1977). Slide in Overconsolidated Clay Below Embankment. *Proc. 9th International Conference on Soil Mechanics and Foundation Engineering*, Tokyo, Vol. 2:95.

Jedelhauser, A. (1970). Durchquerung des Hanges Sonnenberg im Abschnitt Angst-Sissach der Nationalstraße N2. Report of the Schweizerische Gesellschaft für Bodenmechanik und Fundierungstechnik, No. 80.

John, K. W. (1970). Engineering Analysis of Three-Dimensional Stability Problems Ultilizing the Reference Hemisphere. *Proc. 2nd Congress, International Society of Rock Mechanics*, Belgrade, Vol. 3:385-391.

Jubenville, D. M. (1975). A Computer Program for Slope Stability, Modified Bishop Method, Version 1-A BISHOP-1. User's Manual, GESA, University of Colorado Computing Center, Boulder, Co.

Kallstenius, T. (1961) Development of Two Continuous Sounding Methods. *Proc. 5th International Conference on Soil Mechanics and Foundation Engineering*, Vol. 1:475-480.

Kany, M., Stenzel, G., and Neln, H. (1968). Ein Programm zur elektronischen Berechnung der Standsicherheit von Böschungen. Published by Grundbauinstitut der Bayrischen Landesgewerbeanstalt, No. 12.

Kenney, T. C., and Drury, P. (1973). Case Record of the Slope Failure that Initiated the Retrogressive Quick-Clay Landslide at Ullensaker, Norway. *Geotechnique*, Vol. 23, No. 1:33-47.

Kenney, T. C., Pazin, M., and Choi, W. S. (1977). Design of Horizontal Drains for Soil Slopes. *Journal of the Geotechnical Engineering Division*, Vol. 103, GT11, Proc. Paper 13366:1311-1323.

Kézdi, A. (1962). *Erddrucktheorien*. Springer-Verlag, Berlin.

Kézdi, A. (1969). *Handbuch für Bodenmechanik*. Verlag für Bauwesen, Berlin.

Kézdi, A. (1976). *Handbuch der Bodenmechanik, Nr. 4a*. VEB Verlag für Bauwesen, Berlin; Verlag der Ungarischen Akademie der Wissenschaften, Budapest.

Kortüm, G. (1966). *Lehrbuch der Elektrochemie*. Verlag Chemie, Weinheim/Bergstraße.

Krahn, J., and Fredlund, D. G. (1976). Evaluation of the University of Saskatchewan Slope Stability Program. *Proc. Roads and Transportation Association of Canada*, Quebec City, Quebec, Canada.

Krahn, J., Price, V. E., and Morgenstern, N. R. (1971). Slope Stability, Computer Program for Morgenstern-Price Method of Analysis. User's Manual No. 14, University of Alberta, Edmonton, Alberta.

Krey, H. (1926). *Erddruck, Erdwiderstand*. Verlag Wilhelm Ernst & Sohn, Berlin.

Krishnaswamy, V. S. (1980). Geological Aspects of Landslides with Particular Reference to the Himalayan Region. *Proc. International Symposium on Landslides*, New Delhi, Vol. 2:171-185.

Krohn, J. P., and Slosson, J. E. (1976). Landslide Potential in the United States. *California Geology*, Vol. 29, No. 10:224-231.

Kuberan, R., and Tyagi, A. C. (1980). Effect of Water Flow and Earthquake on the Stability of Natural Slopes. *Proc. International Symposium on Landslides*, New Delhi, Vol. 1:303-308.

Ladanyi, B. (1970). The Mechanics of Landslides in Leda Clay: Discussion. *Canadian Geotechnical Journal*, Vol. 7, No. 4:506-507.

Ladanyi, B., Morin, J. P., and Pelchat, C. (1968). Post-Peak Behaviour of Sensitive Clays in Undrained Shear. *Canadian Geotechnical Journal*, Vol. 5, No. 2:59-68.

Ladd, C. C., and Foott, R. (1974). New Design Procedure for Stability of Soft Clays. *Journal of Geotechnical Engineering Division.* Vol. 100, No. GT7:763–786.

Lafleur, J. (1978). Influence de l'eau sur la stabilité des pentes naturelles d'argile. M.A.Sc. Thesis, University of Sherbrooke.

Lambe, T. W. (1967). Stress-Path Method. Proc. ASCE, New York, Vol. 93, SM6:309–311.

La Rochelle, P. (1967). Special Problems in Slope Stability. *Journal of the Soil Mechanics and Foundations Division,* No. SM4:499–526.

La Rochelle, P. (1975). Causes and Mechanism of Landslides in Sensitive Clays With Special Reference to the Quebec Area. *Proc. 4th Guelph Symposium in Geomorphology,* Guelph, Ontario, Canada :173–182.

La Rochelle, P., Chagnon, J. Y., and Lefebvre, G. (1970). Regional Geology and Landslides in the Marine Clay Deposits of Eastern Canada. *Canadian Geotechnical Journal,* Vol. 7, No. 2:145–156.

La Rochelle, P., Lefebvre, G., and Bilodeau, P. M. (1977). Stabilization of a Slide in Saint-Jerome, Lac Saint-Jean. *Canadian Geotechnical Journal,* Vol. 14, No. 3:340–356.

Lauffer, H., Neuhauser, E., and Schober, W. (1971). Der Auftrieb als Ursache von Hangbewegungen bei der Füllung des Gepatschspeichers. *Österreichische Ingenieur-Zeitschrift,* Österr. Ingenieur- und Architektenverein, Wien :101–113.

Law, K. T. (1980). Initial Slides in Leda Clay. *International Symposium on Landslides,* New Delhi, Vol. 1:331–336.

Lawson, A. C. (1908, reprinted 1969). *Landslides.* Carnegie Institute, Washington, D.C., Vol. 1, Part 1:384–401.

Lefebvre, G., Duncan, J. M., and Wilson, E. L. (1973). Three-Dimensional Finite Element Analysis of Dams. *Journal of the Soil Mechanics and Foundations Division,* Vol. 99, No.SM7:495–507.

Lefebvre, G., and Poulin, C. (1979). A New Method of Sampling in Sensitive Clay. *Canadian Geotechnical Journal,* Vol. 16, No. 1:226–233.

Legget, R. F. (1962). *Geology and Engineering.* McGraw-Hill, New York.

Leighton, F. B. (1966). Landslides and Hillside Development. Engineering Geology in Southern California, Association of Engineering Geologists, Special Publication:149–207.

Little, A. L., and Price, V. E. (1958). The Use of an Electronic Computer for Slope Stability Analysis. *Geotechnique,* Vol. 8, No. 3:113–120.

Lo, K. Y. (1965). Stability of Slopes in Anisotropic Soils. *Journal of the Soil Mechanics and Foundations Division,* Vol. 91, No. SM4:85–106.

Lo, K. Y. (1970). The Operational Strength of Fissured Clay. *Geotechnique,* Vol. 20, No. 1:57–74.

Lo, K. Y. (1972). An Approach to the Problem of Progressive Failure. *Canadian Geotechnical Journal,* Vol. 9, No. 4:407–429.

Lo, K. Y., and Morin, J. P. (1972). Strength Anisotropy and Time Effects of Two Sensitive Clays. *Canadian Geotechnical Journal,* Vol. 9, No. 3:261–277.

Løken, T. (1970). Recent Research at the Norwegian Geotechnical Institute Concerning the Influence of Chemical Additions on Quick Clay. *Geologiska Föreningens i Stockholm Förhandlingar,* Vol. 92:133–147.

Lorente de No, C. (1969). Stability With Curvature in Plane View. *Proc. 7th International Conference on Soil Mechanics and Foundation Engineering,* Mexico City, Vol. 2:635–638.

Lowe, J., III (1967). Stability Analysis of Embankments. *Journal of the Soil Mechanics and Foundations Division,* SM4:1–31.

Luceno, A., and Castillo, E. (1980). Evaluation of Variational Methods in Slope Analysis. *Proc. International Symposium on Landslides,* New Delhi, Vol. 1:255–58.

Ludwig, W. (1975). Draukraftwerk Rosegg—St. Jabkob. Geotechnical Report. *Österreichischer Zivil-Ingenieur,* Vol. 28, No. 1:101–105.

Lueder, D. R. (1959). *Aerial Photographic Interpretation: Principles and Applications.* McGraw-Hill, New York.

Lumb, P. (1970). Safety Factors and the Probability Distribution of Soil Strength. *Canadian Geotechnical Journal,* Vol. 7, No. 3:225–242.

Lyndon, A., and Schofield, A. N. (1970). Centrifugal Model Test of a Short-Term Failure in London Clay. *Geotechnique,* Vol. 20:440–442.

Lyndon, A. and Schofield, A. N. (1978). Centrifugal Model Tests of the Lodalen Landslides. *Canadian Geotechnical Journal,* Vol. 15:1–13.

Madhav, M. R., and Ramakrishna, K. S. (1980). *Proc. International Symposium on Landslides,* New Delhi, Vol. 1:267–276.

Maluche, E. (1976). Stützkonstruktionen aus bewehrter Erde. Funktionsprinzip und Anwendungsbeispiele. *Proc. Baugrundtagung Nürnberg,* FRG:653–669.

Mamulea, M. A. (1974). Suffusion and Slaking: Physical Processes Prompting the Mass Movements. *Bull. International Association of Engineering Geologists,* No. 9:63–68 (in French with English summary).

Maruyasu, T., Nakano, T., Nishio, M., Takeda, Y., Kawasaki, T., Kimata, K., Ozarki, Y., and Fuchimoto, M. (1964). Statistical Analysis of Landslides and Related Phenomena on Aerial Photographs. *Journal of Japan Society of Photogrammetry,* Special Vol. 1:93–100.

Massarsch, K. R. (1976). Soil Movements Caused by Pile Driving in Clay. Dept. of Soil and Rock Mechanics, Royal Institute of Technology, Report JoB No. 6.

Massarsch, K. R. (1979). The Stability of Layered Clay Soils. *Proc. 7th European Conference on Soil Mechanics and Foundation Engineering,* Vol. 1:45–51.

Massarsch, K. R., and Broms, B. B. (1976). Lateral Earth Pressure at Rest in Soft Clay. *Journal of the Geotechnical Engineer,* Vol. 102, No. GT8:1041–1047.

Maugeri, M. (1978). Analisi della rottura di un rilevato construtto su un pendio. *Proc. 8th National Geotechnical Conference,* Merano:249–262.

McCauley, M. L. (1976). Microsonic Detection of Landslides. Transportation Research Board, *Transportation Research Record* 581:25–30.

McRoberts, E. C., and Morgenstern, N. R. (1974). Stability of Thawing Slopes. *Canadian Geotechnical Journal,* Vol. 11, No. 4:447–469.

McRoberts, E. C., and Morgenstern, N. R. (1974a). The Stability of Slopes in Frozen Soil: MacKenzie Valley, NWT. *Canadian Geotechnical Journal,* Vol. 11, No. 4:554–573.

Mearns, R., Carney, R., and Forsyth, R. A. (1973). Evaluation of Ion Exchange Landslide Correction Technique. Materials and Research Department, California Division of Highways, Report CA-HY-MR-2116-1-72-39.

Mencl, V. (1966). Mechanics of Landslides With Non-Circular Slip Surfaces With Special Reference to the Vaiont Slide. *Geotechnique* :329–337.

Merriam, R. (1960). Portuguese Bend Landslide, Palos Verdes Hills, California. *Journal of Geology,* Vol. 68, No. 2:140–153.

Meyerhofer, G. G. (1970). Safety Factors in Soil Mechanics. *Canadian Geotechnical Journal,* Vol. 7, No. 4:349–355.

Meyerhof, G. G. (1976). Concepts of Safety in Foundation Engineering Ashore and Off-Shore. *NTH Trondheim, Proc. Boss 76,* Vol. 1:900–912.

Miller, I., and Freund, J. E. (1977). *Probability and Statistics for Engineers.* Prentice-Hall, Englewood Cliffs.

Mitchell, J. K., and Houston, W. N. (1969). Causes of Clay Sensitivity. *Journal of the Soil Mechanics and Foundations Division,* Vol. 95, No. SM3:845–871.

Mitchell, R. J. (1971). On the Yielding and Mechanical Strength of Leda Clays. *Canadian Geotechnical Journal,* Vol. 7:297–312.

Mitchell, R. J., and Markell, A. R. (1974). Flow Sliding in Sensitive Soils. *Canadian Geotechnical Journal,* Vol. 11, No. 1:11–31.

Morgenstern, N. R. (1963). The Limit Equilibrium Method of Slope Stability Analysis. Ph.D. Thesis, University of London.

Morgenstern, N. (1977). Slopes and Excavations in Heavily Overconsolidated Clays. *Proc. 9th International Conference on Soil Mechanics and Foundation Engineering,* Tokyo, Vol. 2:567.

Morgenstern, N. R. (1980). Factors Affecting the Selection of Shear Strength Parameters in Slope Stability Analysis. *Proc. International Symposium on Landslides,* New Delhi, Vol. 2:83–93.

Morgenstern, N. R., and Cruden, D. M. (1977). Description and Classification of Geotechnical Complexities. *Proc. International Symposium on Geotechnics of Structurally Complex Formations,* Capri, Italy.

Morgenstern, N. R., and Price, V. E. (1965). The Analysis of Stability of General Slip Surfaces. *Geotechnique,* Vol. 15, No. 1.

Morgenstern, N. R., and Price, V. E. (1967). A Numerical Method for Solving the Equations of Stability of General Slip Surfaces. *Computer Journal,* Vol. 9, No. 4:388–393.

Morgenstern, N. R., and Sangrey, D. A. (1978). Landslides: Analysis & Control, Methods of Stability Analysis. National Academy of Sciences, Washington D.C., Special Report No. 176:155–191.

Morgenstern, N. R., Blight, G. E., Janbu, N., and Resendiz, D. (1977). Slopes and Excavations. *Proc. 9th International Conference on Soil Mechanics and Foundation Engineering,* Tokyo, Vol. 2:547–604 (see also Krahn *et al.,* 1971, and Krahn and Fredlund, 1976).

Morton, D. M. (1971). Seismically Triggered Landslides in the Area Above the San Fernando Valley. U.S. Geological Survey: The San Fernando, California, Earthquake of February 9, 1971, Professional Paper 733:99–104.

Moum, J. (1966). Elektroosmotisk grunnfursterkning i Ås [Soil Stabilization With Electroosmosis at Ås]. Norwegian Geotechnical Institute, Publ. No. 68:31–38.

Moum, J., Sopp, O. I., and Løken, T. (1968). Stabilization of Undisturbed Quick Clay by Salt Wells. Norwegian Geotechnical Institute, Publ. No. 81.

Müller, L. (1964). The Rock Slide in the Vajont Valley. *Felsmechanik und Ingenieurgeologie,* Vol. 1:148–212.

Nakamura, H. (1980). Displacement Analysis of Sarukuyoji Landslide on the Basis of a Consolidation Theory. *Proc. International Symposium on Landslides,* New Delhi, Vol. 1:309–312.

Nakase, A. (1970). Stability of Low Embankment on Cohesive Soil Stratum. *Soils and Foundations,* Vol. 10, No. 4:39–64.

Nakayama, Y. (1980). A Study of Stability Analysis on Landslides. *Proc. International Symposium on Landslides,* New Delhi, Vol. 1:177–180.

Nandakumaran, P. (1980). On Seismic Stability Problems of Some Geotechnical Constructions. *Proc. International Symposium on Landslides,* New Delhi, Vol. 1:79–84.

Nandakumaran, P., Prakash, S., and Singh, A. (1980). Factor of Safety of Slopes Against Slumps. *Proc. International Symposium on Landslides,* New Delhi, Vol. 1:289–292.

Natarajan, T. K., and Gupta, S. C. (1980). Techniques of Erosion Control for Surficial Landslides. *Proc. International Symposium on Landslides,* New Delhi, Vol. 1:413–418.

Nelson, J. D., and Thompson, E. G. (1977). A Theory of Creep Failure in Overconsolidated Clay. *Journal of the Geotechnical Engineering Division,* Vol. 103, No. GT 11:1281–1294.

Nemcok, A., Pasek, J., and Rybar, J. (1972). Classification of Landslides and Other Mass Movements. *Rock Mechanics,* Vol. 4, No. 2:71–78.

Newland, P. L. (1967). A Note on the Slices Method of Stability Analysis. *Proc. 5th Australia–New Zealand Conference on Soil Mechanics and Foundation Engineering,* University of Auckland.

Newmark, N. M. (1965). Effects of Earthquakes on Dams and Embankments. *Geotechnique,* Vol. 15, No. 2:139–159.

Nishida, Y., Yagi, N., and Futaki, M. (1979). The Theoretical Analysis of the Pore Pressures Due to Rain Water Permeation in the Ground. *Proc. 6th Asian Regional Conference on Soil Mechanics and Foundation Engineering,* Singapore, Vol. 1:241–244.

Nonveiller, E. (1965). The Stability Analysis of Slopes with a Slip Surface of General Shape. *Proc. 6th International Conference on Soil Mechanics and Foundation Engineering,* Montreal, Vol. 2: 522–526.

Oba, N., Tomita, K., and Yamamoto, M. (1980). Physico-Chemical Features of the 'Shirasu' Pumice Flow Deposits and Its Related Landslides in South Kyushu, Japan. *Proc. International Symposium on Landslides,* New Delhi, Vol. 1:13–18.

Okubo, S., Yamazaki, S., and Okuzono, S. (1980). On the Stability and Weathering of Cut Slopes. *Proc. International Symposium on Landslides,* New Delhi, Vol. 1:195–198.

Orrje, O., and Broms, B. (1968). Effect of Pile Driving on Pore Pressures and Shear Strength. *Journal of the Soil Mechanics and Foundations Division,* Vol. 93, SM 5, Part I:59–73.

Osterman, J. (1964). Studies on the Properties and Formation of Quick Clays. Clays and Clay Mineral Monograph No. 19:87–108. Earth Science Series, Pergamon Press, Oxford.

Oyagi, N. (1980). Landslide Triggered by the 1978 Izu-Oshima Earthquake. *Proc. International Symposium on Landslides,* New Delhi, Vol. 1:89–94.

Oyagi, N., and Nakamura, H. (1977). *9th International Conference on Soil Mechanics and Foundation Engineering,* Tokyo, Guide Book for Excursions of Landslides in Central Japan. The Japan Society of Landslides:16–19.

Oyagi, N., Fujita, T., and Nakamura, H. (1977). The Sarukuyoji Landslide. *The Japan Society of Landslides,* Report.

Patel, A. N., and Mehta, H. S. (1980). Application of Digital Techniques in Identifying Landslide-Prone Areas. *Proc. International Conference on Landslides,* New Delhi, Vol. 1:25–28.

Peck, R. B (1967). Stability of Natural Slopes. *Journal of the Soil Mechanics and Foundations Division,* Vol. 93, SM4:403–417.

Peck, R. B. (1967a). Observation and Instrumentation: Some Elementary Considerations. Seminar on Field Observations in Foundation Design and Construction, ASCE, New York. See *Highway Focus,* U.S. Department of Transportation, FHA, Vol. 4, No. 2:1–6.

Peck, R. B. (1969). Advantages and Limitations of the Observation Method in Applied Soil Mechanics. *Geotechnique,* Vol. 19, No. 2.

Peck, R. B., Hanson, W. E., and Thronbum, T. H. (1973). *Foundation Engineering.* Wiley, New York.

Petar, L. (1980). Engineering—Geological Factors in Designing Landslide Control. *Proc. International Symposium on Landslides,* New Delhi, Vol. 1:407–412.

Petterson, K. E. (1955). The Early History of Circular Sliding Surfaces. *Geotechnique,* Vol. 5:275–296.

Pilot, G. (1970). The Stability of Slopes. *Construction,* Paris; 1970—Vol. 25, No. 10:338–342; 1971—Vol. 26, No. 7/8:274–280; 1972—Vol. 27, No. 2:61–69.

Pilot G., and Moreau, M. (1973) *La Stabilité des Remblais Sur Sols Mous.* Eyrolles, Paris.

Pincent, B., Queroi, D., Pilot. G., and Baroux, R. (1977). Centrifugal Model Test of Failure of a Cut. *Proc. 9th International Conference on Soil Mechanics and Foundation Engineering,* Vol. 1:251–254.

Pötscher, R. (1977). Gipszellenversuche. Publication of the Graz Technical University, Department of Soil Mechanics, Rock Mechanics, and Geotechnical Engineering, No. 2:27–31.

Prakash, S., Ranjan, G., Saran, S., Singh, B., and Ramasay, G. (1980). Evaluation of Stability of Slopes in Himalayan Region. *Proc. International Symposium on Landslides,* New Delhi, Vol. 1:165–168.

Prinzl, F. (1977). Elektrische Spannungspotentialmessungen im Tunnel Las Planas, Frankreich (N 770). Publication of the Graz Technical University, Department of Soil Mechanics, Rock Mechanics and Geotechnical Engineering, No. 4:50–57.

Prior, D. B., Stephens, N., and Archer, D. R. (1968). Composite Mudflows on the Antrim Coast of Northeast Ireland. *Geografiska Annaler,* Vol. 50A, No. 2:65–78.

Prior, D. B., Stephens, N., and Douglas, G. R. (1970). Some Examples of Modern Debris Flows in Northeast Ireland. *Zeitschrift für Geomorphologie,* Vol. 14, No. 3:276–288.

Pusch, R. (1970). Microstructural Changes in Soft Quick Clay at Failure. *Canadian Geotechnical Journal,* Vol. 7, No. 1:1–7.

Radbruch-Hall, D. H., Colton, R. B., Davies, W. E., Skipp, B. A., Lucchitta, I., and Varnes, D. J. (1976). Preliminary Landslide Overview Map of the Coterminous United States. U.S. Geological Survey, Miscellaneous Field Studies, Map MF-771.

Ramamurthy, T., Narayan, C. G. P., and Bhat Kar, V. P. (1977). Variational Methods for the Slope Stability Analysis. *Proc. 9th International Conference on Soil Mechanics and Foundation Engineering,* Tokyo, Vol. 2:139–142.

Raschka, H. (1912). Die Rutschungen in dem Abschnitt Ziersdorf–Eggenburg der Kaiser-Franz-Josefs-Bahn (Hauptstrecke). *Zeitschrift des Österreichischen Ingenieur- und Architektenvereines,* September 1912.

Redlich, K. R., Terzaghi, K., and Kampe, K. (1929). Ingenieurgeologie. Springer, Berlin.

Rendulic, L. (1940). Gleitflächen, Prüfflächen, Erddruck. *Die Bautechnik, Fachzeitschrift für das gesamte Bauingenieurwesen,* No. 13/14.

Rib, H. T., and Liang, T. (1978). Landslides Analysis and Control, Recognition and Identification. National Academy of Sciences, Washington D.C., Special Report No. 176:34–80.

Ribbert, H. (1980). System Ribbert. Company-Internal Report, Bad Aiblingen, Oberbayern, FRG.

Rice, R. M., Corbett, E. S., and Bailey, R. G. (1969). Soil Slips Related to Vegetation, Topography, and Soil in Southern California. *Water Resources Research,* Vol. 5, No. 3:647–659.

Rico, A. (1980). Some Remarks on the Stability Analysis Methods. *Proc. International Symposium on Landslides,* New Delhi, Vol. 1:277—280.

Rosenqvist, I. Th. (1975). Clay Mineralogy Applied to Mechanics of Landslides. *Geologia Applicata e Idrogeologia,* Vol. 10, Part II:21–32.

Rosenqvist, I. Th. (1977). A General Theory for Quick Clay Properties. *Proc. 3rd European Clay Conference,* Oslo:215–228.

Rowe, P. W. (1972). Embankments on Soft Alluvial Ground. *Quarterly Journal of Engineering Geology,* Vol. 5:127–149.

Ruenkreirergsa, T., and Chinpongsanond, P. (1980). Geological and Seismological Aspects of Landslides, *Proc. International Symposium on Landslides,* New Delhi, Vol. 1:85–88.

Rüsch, H. (1971). Sicherheitsfragen, *Betonkalender* 1971:640–729.Wilhelm Ernst & Sohn, Berlin.

Saito, M. (1970). Estimation of the Rupture Life of Soil Based on the Shape of the Creep Curve. *Proc. 5th International Congress on Rheology,* Vol. 2:559–567.

Saito, M. (1980). Semi-Logarithmic Representation for Forecasting Slope Failure. *Proc. International Symposium on Landslides,* New Delhi, Vol. 1:321–324.

Sangrey, D. A. (1972). Naturally Cemented Sensitive Soils. *Geotechnique,* Vol. 22, No. 1:138–152.

Sangrey, D. A. (1972a). Changes in Strength of Soil Under Earthquake and Other Repeated Loading. *Proc. 1st Canadian Conference on Earthquake Engineering,* University of British Columbia, Vancouver:82–96.

Sassa, K., Takei, A., and Kobashi, S. (1980). Landslides Triggered by Vertical Subsidences. *Proc. International Symposium on Landslides,* New Delhi, Vol. 1:49–54.

Sassa, K., Takei, A., and Kobashi, S. (1980a). Considerations of Vertical Subsidences as a Factor Influencing Slope Instability. *Proc. International Symposium on Landslides,* New Delhi, Vol. 1:293–298.

Schiffmann, R. L. (1977). Computers in Soil Mechanics: Present and Future. *Proc. of the Speciality Session on Computers in Soil Mechanics: Present and Future; 9th International Conference on Soil Mechanics and Foundation Engineering,* Tokyo; MAA Publishing Co.

Schindler, J. (1969). Untersuchungen der Wirkungsweise einer mehrfach verankerten Wand in kohäsionslosem Erdmaterial. Publication of the Versuchsanstalt für Wasserbau und Erdbau, Eidgenössische Technische Hochschule Zürich, No. 83.

Schmertman, J. H. (1976). The Shear Behaviour of Soil with Constant Structure. Laurits Bjerrums Memorial Volume, NGI, Oslo.

Schmid, G. (1951). Zur Elektrotechnik feinporiger Kapillarsysteme. II. Elektroosmose. *Electrochemie,* No. 55:229–237.

Schmid, G., and Schwarz, H. (1951). Zur Elektrochemie feinporiger Kapillarsysteme. III. Elektrische Leitfähigkeit. *Elektrochemie* No. 55:295–307.

Schofield, A. N. (1976). Use of Centrifugal Model Testing to Assess Slope Stability. Cambridge University Engineering Department Report No. CUED/C-Soils/TR-30.

Schubert, K. (1972). *Böschungen, Dämme, Halden, Klippen.* VEB Deutscher Verlag für Grundstoffindustrie, Leipzig.

Schultze, E., and Muhs, H. (1967). *Bodenuntersuchungen für Ingenieurbauten.* Springer-Verlag, Berlin-Heidelberg-New York.

Schuster, R. L. (1978). Landslides: Analysis and Control, Introduction. National Academy of Sciences, Washington, D.C., Special Report No. 176:1–10.

Schweizer, R. J., and Wright, S. G. (1974). A Survey and Evaluation of Remedial Measures for Earth Slope Stabilization. Center for Highway Research, University of Texas at Austin, Research Report 161-2F.

Schwinn, K. H., and Hardt, G. (1968). Zur Standsicherheit von Böschungen nach der Erddruckregel. Donau-Europäische Konferenz über Bodenmechanik im Straßenbau, Österreichischer Ingenieur- und Architektenverein, Vol. 1:175–178.

Scott, R. F. (1980). Slope Stability Studies in the Centrifuge. *Proc. International Symposium on Landslides,* New Delhi, Vol. 2:67–82.

Seed, H. B. (1968). Landslides During Earthquakes Due to Soil Liquefaction. *Journal of the Soil Mechanics and Foundations Division,* Vol. 94, SM5:1055–1122.

Seed, H. B., and Peacock, W. H. (1971). Test Procedures for Measuring Soil Liquefaction Characteristics. *Journal of the Soil Mechanics and Foundations Division,* Vol. 97, SM8:1099–1119.

Seed, H. B., and Wilson, S. D. (1967). The Turnagain Heights Landslide, Anchorage, Alaska. *Journal of the Soil Mechanics and Foundations Division,* Vol. 93, SM4:325–353.

Seltenhammer, U. (1968). Ankermauer an der Brennerautobahn. *Österreichische Ingenieur-Zeitung,* Vol. 11, No. 6.

Siegel, R. A. (1975). Computer Analysis of General Slope Stability Problems. Joint Highway Research Project No. C-36-36K, Engineering Experimental Station, Purdue University, West Lafayette, Indiana, JHRP-75-8:1–112.

Sima, H., and Izutani, Y. (1980). Landslide Movement Expected from Observations of Microtremors. *Proc. International Symposium on Landslides,* New Delhi, Vol. 1:375–382.

Singh, B. (1980). Critical Evaluation of Theoretical Approaches to Landslide Studies. *Proc. International Symposium on Landslides,* New Delhi, Vol. 2:23–25.

Singh, B., and Ramasamy, G. (1979). Back-Analysis of Natural Slopes for Strength Parameters. *Proc. International Conference on Computer Applications in Civil Engineering,* University of Roorkee.

Singh, J. et al. (1975). Selection and Design of Structures for Protection and Prevention of Landslides and Toe Erosion. *Proc. Seminar on Landslides and Toe Erosion, with special reference to Himalayan Region,* Gangtok:275.

Singh, R., Henkel, D. J., and Sangrey, D. A. (1973). Shear and K_0 Swelling of Overconsolidated Clay. *Proc. 8th International Conference on Soil Mechanics and Foundation Engineering,* Moscow, Vol. 1, Part 2:367–376.

Sinha, B. N. (1975). An Engineering Geologic Approach to Landslides and Slope Failures. *Proc. Seminar on Landslides and Toe Erosion, with special reference to the Himalayan Region,* Gangtok:54.

Siniscalchi, C. (1965). Remarks on the Stability of Analysis of Earth Slopes. *Proc. 6th International Conference on Soil Mechanics and Foundation Engineering, Montreal.* University of Toronto Press, Vol. 2:556–560.

Sior, G. (1961). Grenzgleichgewicht und Sicherheitsgrad von Böschungen in einfachen Fällen. *Der Bauingenieur, Zeitschrift für das gesamte Bauwesen,* No. 11:423–425.

Sior, G. (1968). Die Kinematik als Grundlage der Standsicherheitsberechnungen von Böschungen. *Proc. Baugrundtagung Hamburg,* Deutsche Gesellschaft für Erd- und Grundbau e.V., Essen :105–117.

Siva Reddy, A., and Venkatakrishna, R. (1980). Stability of Slopes With Foundation Loads. *Proc. International Symposium on Landslides,* New Delhi, Vol. 1:285–293.

Skempton, A. W. (1953). The Colloidal Activity of Clays, *Proc. 3rd International Conference on Soil Mechanics and Foundation Engineering,* Zürich.

Skempton, A. W. (1954). The Pore-Pressure Coefficients A and B. *Geotechnique,* Vol. 4, No. 4:143–147.

Skempton, A. W. (1961). Horizontal Stress in an Over-Consolidated Eocene Clay. *Proc. 5th International Conference on Soil Mechanics and Foundation Engineering,* Paris, Vol. 1:351–357.

Skempton, A. W. (1964). Long-Term Stability of Clay Slopes. *Geotechnique,* Vol. 14:77–100.

Skempton, A. W. (1969). Slope Stability of Cuttings in Brown London Clay. Special Lecture. *Proc. 7th International Conference on Soil Mechanics and Foundation Engineering,* Mexico :26–29.

Skempton, A. W. (1970). First-Time Slides in Over-Consolidated Clays. *Geotechnique,* Vol. 20, No. 3:320–324.

Skempton, A. W. (1977a). Slope Stability of Cuttings in Brown London Clay. *Proc. 9th International Conference on Soil Mechanics and Foundation Engineering,* Tokyo, Vol. 3:261–270.

Skempton, A. W. (1977b). Slope Stability of Cuttings in Brown London Clay. *Proc. 9th International Conference on Soil Mechanics and Foundation Engineering,* Tokyo, Special Lectures Vol.:25–33.

Skempton, A. W., and Brown, J. D. (1961). A Landslide in Boulder Clay at Selset, Yorkshire. *Geotechnique,* Vol. 11, No. 4:280–293.

Skempton, A. W., and Golder, H. Q. (1948). Practical Examples of the $\phi = 0$ Analyses of Stability of Clays. *Proc. 2nd International Conference on Soil Mechanics and Foundation Engineering,* Rotterdam, Vol. 2:63–70.

Skempton, A. W., and Hutchison, J. N. (1969). Stability of Natural Slopes and Embankment Foundations. *Proc. 7th International Conference on Soil Mechanics and Foundation Engineering,* Mexico City, State-of-the-Art Vol.:291–340.

Skempton, A. W., and La Rochelle, P. (1965). The Bradwell Slip: A Short-Term Failure in London Clay. *Geotechnique,* Vol. 15, No. 3:221–242.

Skempton, A. W., and Northey, R. D. (1952). The Sensitivity of Clays. *Geotechnique,* Vol. 3, No. 1:30–53.

Skempton, A. W., and Petley, D. J. (1967). The Strength Along Structural Discontinuities in Stiff Clays. *Proc. Geotechnical Conference on Shear Strength Properties of Natural Soils and Rocks,* Norwegian Geotechnical Institute, Oslo, Vol. 2:29–46.

Smith, I. M., and Hobbs, R. (1974). Finite Element Analysis of Centrifuged and Built-Up Slopes. *Geotechnique,* Vol. 24:531–559.

Smoltczyk, U. (1975). Anmerkungen zum Gleitkreisverfahren, Festschrift zum 70. Geburtstag von Herrn Professor Dr. Ing. H. Lorenz. Berlin: Universitätsbibliothek der Technischen Universität, Berlin.

Söderblom, R. (1960). Aspects of Some Problems of Geotechnical Chemistry. *Geologiska Föreningens Stockholm Förhandlingar,* Part II, Vol. 82, No. 3:367–381.

Söderblom, R. (1966). Chemical Aspects of Quick Clay Formation. *Engineering Geology,* Vol. 6, No. 1:415–431.

Söderblom, R. (1969). Salt in Swedish Clays and Its Importance for Quick Clay Formation. *Proc. Swedish Geotechnical Institute,* No. 22.

Söderblom, R. (1974a). A New Approach to the Classification of Quick Clays. *Swedish Geotechnical Institute, Reprints and Preliminary Reports,* No. 55:1–17.

Söderblom, R. (1974b). Organic Matter in Swedish Clays and Its Importance for Quick Clay Formation. *Swedish Geotechnical Institute,* No. 26.

Som, N. (1980). Properties of Materials Relevant to Landslide Studies. *International Symposium on Landslides,* New Delhi, Vol. 2:7–11.

Sommer, H. (1977). Kriechender Hang im Tertiärton—Bodenmechanische Probleme. *Proc. 9th International Conference on Soil Mechanics and Foundation Engineering,* Tokyo, Specialty Session 10 (unpublished).

Soos, P. (1970). Standsicherheit von Böschungen. VDI Berichte No. 142:45–53.

Sowers, G. F., and Royster, D. L. (1978) Field Investigations. National Academy of Sciences, Washington, D.C., Special Report 176:81–111.

Spencer, E. (1968). Effect of Tension on Stability of Embankments. *Journal of the Soil Mechanics and Foundations Division,* SM5:1159–1173.

Spencer, E. (1969). Circular and Logarithmic Spiral Slip Surfaces. *Journal of the Soil Mechanics and Foundations Division,* SM95:227–234.

Stocker, M., and Bauer, K. (1976). Bodenvernagelung. *Proc. BaugrundTagung Nürnberg* : 639–652.

Suklje, L., and Vidmar, S. (1961). A Landslide Due to Long-Term Creeps. *Proc. 5th International Conference on Soil Mechanics and Foundation Engineering,* Vol. 2:727.

Swanston, D. N. (1974). Slope Stability Problems Associated With Timber Harvesting in Mountainous Regions of the Western United States. U.S. Forest Service, General Technical Report PNW-21.

Taylor, D. W. (1937). Stability of Earth Slopes. Contributions to Soil Mechanics, 1925–40, BSCE, 337–386.

Ter-Stepanian, G. (1963). On the Long-Term Stability of Slopes. Norwegian Geotechnical Institute, Publ. 52.

Ter-Stepanian, G. (1967). The Use of Observations of Slope Deformation for Analysis of Mechanism of Landslides: Problems of Geomechanics. Armenian SSR Academy of Sciences, No. 1:32–51.

Ter-Stepanian, G. (1974). Depth Creep of Slopes. *Bull. International Society of Engineering Geologists,* No. 9:97–102.

Ter-Stepanian, G. (1977). Behaviour of Clays Close to Failure. *Proc. 9th International Conference on Soil Mechanics and Foundation Engineering,* Tokyo, Vol. 1:327–328.

Ter-Stepanian, G. (1980). Measuring Displacements of Wooded Landslides with Trilateral Signs. *Proc. International Symposium on Landslides,* New Delhi, Vol. 1:355–360.

Ter-Stepanian, G. (1980). Creep of Natural Slopes and Cuttings, *Proc. International Symposium on Landslides,* New Delhi, Vol. 2:95–108.

Ter-Stepanian, G., and Goldstein, M. N. (1969). Multi-Storied Landslides and Strength of Soft Clays. *Proc. 7th International Conference on Soil Mechanics and Foundation Engineering,* Mexico City, Vol. 2:693–700.

Terzaghi, K. (1931). Earth Slips and Subsidences from Underground Erosion. *Engineering News-Record,* Vol. 107:90–92.

Terzaghi, K. (1936). The Shearing Resistance of Saturated Soils. *Proc. 1st International Conference on Soil Mechanics and Foundation Engineering,* Cambridge, Mass., Vol. 1:54–56.

Terzaghi, K. (1936a). Critical Height and Factor of Safety of Slopes. *Proc. 1st International Conference on Soil Mechanics and Foundation Engineering,* Cambridge, Mass., Vol. 1:156.

Terzaghi, K. (1944). *Theoretical Soil Mechanics.* John Wiley & Sons, New York.

Terzaghi, K. (1950). Mechanism of Landslide. *Geological Society of America, Engineering Geology,* Vol. 83.

Terzaghi, K. (1953). Discussion on Stability and Deformation of Slopes and Earth Dams, Etc. *Proc. 3rd International Conference on Soil Mechanics and Foundation Engineering,* Zürich, Vol. 3:217–218.

Terzaghi, K. (1961). Discussion. *Proc. 5th International Conference on Soil Mechanics and Foundation Engineering,* Paris, Vol 111:144.

Terzaghi, K. (1962). Stability of Steep Slopes on Hard Unweathered Rock. *Geotechnique,* Vol. 12, No. 4:251–270.

Terzaghi, K., and Peck, R. (1961). *Bodenmechanik in der Baupraxis.* Springer, Berlin-Göttingen-Heidelberg.

Terzaghi, K., and Peck, R. (1967). *Soil Mechanics in Engineering Practice.* John Wiley & Sons, New York.

Trollope, D. H. (1957). The Systematic Arching Theory Applied to the Stability Analysis of Embankments. *Proc. 4th International Conference on Soil Mechanics and Foundation Engineering,* London, Butterworths Scientific Publications, Vol. 2:382–388.

Trollope, D. H. (1973). Sequential Failure in Strain-Softening Soils. *Proc. 8th International Conference on Soil Mechanics and Foundation Engineering,* Moscow, Vol. 2, Part 2:227–232.

Torrance, J. K. (1975). On the Role of Chemistry in the Developments and Behaviour of the Sensitive Marine Clays of Canada and Scandinavia. *Canadian Geotechnical Journal,* Vol. 12, No. 3:326–335.

Torstensson, B.-A. (1975). Pore Pressure Sounding Instrument. Discussion. *ASCE Specialty Conference on in situ Measurement of Soil Properties,* North Carolina State University, Raleigh, Vol. 2:48–54.

Uriel, R. S. (1969). Design of Loose-Fill Dam Slopes by the Method of Characteristics. *Proc. 7th International Conference on Soil Mechanics and Foundation Engineering,* Mexico City, Vol. 2:387–395.

Varnes, D. J. (1972). A Classification of Landslides. *Symposium on Landslides Control,* Japan Society on Landslides, Kyoto :65–79.

Varnes, D. J. (1978). Slope Movement Types and Processes. Landslides: Analysis and Control. National Academy of Sciences, Washington, D.C., Special Report 176:11–33.

Vaughan, P. R., and Walbancke, H. J. (1973). Pore Pressure Changes and the Delayed Failure of Cutting Slopes in Overconsolidated Clay. *Geotechnique,* Vol. 23, No. 4:531–539.

Veder, Ch. (1957). Considerazioni sulla possibilità che fenomeni ellettro-osmotici siano all'origine della formazione di particolari tipi di frane. *Geotecnica,* Vol. 5:3–11.

Veder, Ch. (1963). Die Bedeutung natürlicher elektrischer Felder für Elektroosmose and Elektrokataphorese. *Der Bauingenieur,* No. 10:378–388.

Veder, Ch. (1964). Stabilisierung von Rutschungen die durch Elektroosmose entstanden sind, mittel Aufhebung natürlicher elektrischer Felder. Baugrundtagung, Berlin :91–113.

Veder, Ch. (1966). Moderne Rutschungssicherung. Presentation before the Meeting of the Forschungsgesellschaft für das Straßenwesen, Österreichischer Ingenieur- und Architektenverein, Wien.

Veder, Ch. (1968). Bodenstabilisierung durch Ausschalten von Grenzflächenerscheinungen. Springer, Wien-New York. *Felsmechanik und Ingenieurgeologie,* Supp. IV:9–24.

Veder, Ch. (1968a). Landslides Stabilized by Using Electrodes. *Ground Engineering,* January 1968.

Veder, Ch. (1972a). Grenzflächenerscheinungen in der Bodenmechanik. *Der Bauingenieur.* No. 3:80–89.

Veder, Ch. (1972b). Origin of Groundwater, Particularly in Landslides. Lecture at the Japan Landslide Commission, Kyoto.

Veder, Ch. (1972c). Phenomena of the Contact of Soil Mechanics. *Proc. 1st International Symposium on Landslide Control,* Japan Society of Landslide, Kyoto and Tokyo :143–162. (in English and Japanese).

Veder, Ch. (1973). The Phenomenon of Contact Zones of Soil Mechanics. *Ground Engineering,* Vol. 6, No. 5.

Veder, Ch., and Finzi, D. (1962). Stabilizzazione di una frana mediante infusione di elettrodi nel piano di scivolamento. *Geotecnica,* Vol. 5:2–8.

Veder, Ch., and Hilbert, F. (1980). Electroosmotical Stabilization of Slopes. *Proc. International Symposium on Landslides,* New Delhi, Vol. 2:155–169.

Vutsel, V. I., and Shcherbina, V. I . (1973). Investigations of RockFill Dam Core Cracking by Centrifugal Modeling. *Proc. 8th International Conference on Soil Mechanics and Foundation Engineering,* Vol. 4. 1:209–210.

Wagner, H. (1974). Grenzflächenrutschungen. Thesis, Department of Soil Mechanics, Rock Mechanics, and Geotechnical Engineering, Graz Technical University.

Watari, M. (1973). For the Relationship Between Large-Scale Earthworks and the Occurrence of Landslides in Natural Slope. *Proc. 8th International Conference on Soil Mechanics and Foundation Engineering,* Moscow, Specialty Session 1:347.

Wenzel, K., and Fenz, M. (1970). Die Luegbrücke im Zuge der Brennerautobahn. *Österreichische Ingenieurzeitung,* Vol. 13, No. 1:67–76.

Whitman, R. V., and Bailey, W. A. (1967). Use of Computers for Slope Stability Analysis. *Journal of the Soil Mechanics and Foundations Division,* SM4:474–497. (See also *Proc. 6th International Conference of Soil Mechanics and Foundations Engineering,* Montreal, 1966, Vol. 3:449–450.)

Wilson, S. D. (1970). Observational Data on Ground Movements Related to Slope Instability. *Journal of the Soil Mechanics and Foundations Division,* Vol. 96, SM5:1521–1544.

Wilson, S. D. (1974). Landslide Instrumentation for the Minneapolis Freeway. Transportation Research Board, Record 482:30–42.

Wilson, S. D., and Mikkelsen, P. E. (1978). Field Instrumentation. Landslides: Analysis and Control. National Academy of Sciences, Washington, D.C., Special Report 176:112–138.

Winterkorn, H. F., and Fang, H. Y. (1975). *Foundation Engineering Handbook.* Van Nostrand Reinhold, New York.

Wissa, A. E. Z., Martin, R. T., and Garlanger, J. E. (1975). The Piezometer Probe. ASCE Specialty Conference on in situ Measurement of Soil Properties, North Carolina State University, Raleigh, Vol. 1:536–545.

Wu, T. H. (1974). Uncertainty, Safety, and Decision in Soil Engineering. *Journal of the Geotechnical Engineering Division,* Vol. 100, GT3:329.

Wu, T. H., and Sangrey, D. A. (1978). Strength Properties and Their Measurement. Landslides: Analysis and Control, National Academy of Sciences, Washington, D.C., Special Report 176:139–154.

Wysokinski, L. (1980). Forecasting of Slope Deformation—A New Criterion. *Proc. International Symposium on Landslides,* New Delhi, Vol. 1:325–330.

Wysokinski, L., and Ostaficzuk, S. (1980). Landslide Investigation and Preservation of the Plock City Escarpment. *Proc. International Symposium on Landslides,* New Delhi, Vol. 1:425–430.

Youd, T. L. (1973) Liquefaction, Flow, and Associated Ground Failure. U.S. Geological Survey, Circular 688.

Yudhbir, (1980). Rebound Characteristics of Overconsolidated Clays and Clay Shales. *Proc. International Symposium on Landslides,* New Delhi, Vol. 1:119–128.

Yudhbir, (1980). Field Studies Relating to Natural Slopes and Cuttings. *Proc. International Symposium on Landslides,* New Delhi, Vol. 2:13–21.

Zaruba, Q., and Mencl, V. (1969). *Landslides and Their Control.* Elsevier, Amsterdam-London-New York.

Index